MINISTRY OF
AGRICULTURE, FISHERIES AND FOOD

Diseases of Bees

Bulletin 100

LONDON
HER MAJESTY'S STATIONERY OFFICE

ISBN 0 11 240400 6

Foreword

GOOD beekeeping requires not only a knowledge of the basic principles of colony management but also an acquaintance with the nature of the various diseases which may infect the honeybee either in its larval or its adult form. Early diagnosis and the prompt application of control measures once the cause of the trouble has been ascertained are doubly important—to avoid unnecessary losses among the beekeeper's own stock and to prevent the spread of disease to other apiaries.

This fourth edition has been revised by the Ministry's Bee Group under the chairmanship of Mr. F. A. Richards, County Beekeeping Instructor, Norfolk. Special thanks are due to Dr. L. Bailey, Rothamsted Experimental Station for his helpful criticisms and comments on the text during the process of revision.

Ministry of Agriculture,
Fisheries and Food
March, 1969

Contents

Brood Diseases

THE brood of the honeybee, during its development from the egg to the transformation of the pupa, is subject to a variety of diseases, all of which have a weakening effect on the colony as a whole. The degree of weakening ranges from the loss of a small proportion of the total population of the hive, in a mild attack of chalk brood, for example, to the complete collapse and eventual death of the colony in the case of an infection with American foul brood. This disease in particular is responsible for considerable losses to beekeepers every year, and a single neglected case may become a source of infection for any other bees which are kept in the neighbourhood.

It is, therefore, very desirable that every beekeeper should have a working knowledge of the main features of the more common brood diseases and that he should at least be able to distinguish between healthy and diseased brood, even if he cannot make a diagnosis of the precise cause of the trouble without expert help.

A careful inspection of the brood should be made during all manipulations involving the removal of combs from the hive, or their rearrangement, A casual glance at the *amount* of brood present is not enough and both the sealed and the unsealed cells should receive their share of attention. If anything suspicious is noticed it should be investigated at once, since failure to do so may result in the spread of disease to other colonies in the same apiary or to neighbouring apiaries. It should be pointed out, however, that the first signs of a brood disease are not easily detected except at close quarters, especially if the combs are old and dark in colour, and beekeepers who require spectacles for reading purposes should always use them under the bee-veil when making an examination of the brood chamber.

FOUL BROOD

The term 'Foul Brood' includes two bacterial diseases of the honeybee, one known as American foul brood, and the other as European foul brood. The names bear no relation either to the origin or to the geographical distribution of the two diseases; they came into use after research work in America had shown that the bacteriology of the 'Foul Brood' then being investigated there differed from that which had been described from another case of 'Foul Brood' investigated some years earlier in England.

American foul brood is the most widespread and the most destructive of the brood diseases which occur in Great Britian. Cases have been reported from all the counties in England and Wales (including the Isle of Wight), from parts of Scotland, and also from Northern Ireland and Eire.

European foul brood is less common, but where it does occur infection often spreads unseeen through the apiary, and for this reason may prove difficult to eradicate even when control measures are applied promptly.

AMERICAN FOUL BROOD

Cause

American foul brood is caused by a microscopic spore-forming organism known as *Bacillus larvae*. Spores from some previously existing source of infection, such as contaminated honey robbed from another colony already weakened by the disease (see under Spread, page 4), become mixed with the brood food fed to young larvae by the nurse bees. The spores germinate within the body of the larva and produce bacteria, which then grow and multiply at a great rate, feeding at the expense of the tissues of the larva itself. Soon after the larva has been sealed over in its cell by the bees working on the comb, it collapses and dies. When this happens the food supply of the bacteria is no longer maintained, and their growth and multiplication cease. Each bacterium then transforms itself back into the spore stage. The spores so produced, now many times more numerous than those which caused the larva to become infected in the first place, remain dormant until some of them, in their turn, are picked up by house-cleaning bees attempting to clean out the cell containing the dead remains of the larva. By this means they are distributed throughout the hive and complete their life cycle by being mixed with the brood food of another larva. The whole process then starts again and is repeated indefinitely. More and more larvae become infected, the proportion of the brood which emerges gradually becomes less and less, and sooner or later the colony dies out from sheer lack of enough bees to keep it going. Drone larvae are susceptible to infection as well as worker larvae; occasionally a queen larva may be found to be affected, though the bees seldom attempt to produce queen cells once the disease has spread through the brood nest and has begun to reduce the strength of the colony.

The spores are very resistant to exposure, to extremes of heat and cold, and to disinfectants. They will retain their powers of germination for many years if left undisturbed in old combs kept in store, or in derelict hives, skeps or boxes. To see the spores it is necessary to spread a thin smear of the dead larval remains on to a glass slide and to stain the smear with a suitable dye for examination under a good microscope giving a magnification of about 1,000 diameters.

Symptoms

The collapse of the infected larva takes place within the sealed cell, after the cocoon has been spun, and is, therefore, not to be seen merely by looking at the surface of the comb. Associated with this collapse, however, are changes in the appearance of the cell capping, and these changes are of primary importance as a first indication of the presence of abnormal brood in the hive (Plate I). The capping becomes moist and darkens in colour, and as the larva continues to shrink the capping is drawn down into the mouth of the cell, its originally slightly domed shape becoming reversed. The worker bees nibble holes in the sunken capping (Plate II) and eventually remove it altogether, leaving an open cell containing the remains of the dead larva (see figure on page 3 and Plate III).

The collapse of the larva is accompanied by a change in colour from the normal pearly-white of healthy brood to a creamy-brown, light at first, then becoming darker. The consistency of the larval remains is very slimy at this stage and if a match-stick is thrust through the sunken capping into the cell, twisted round and then withdrawn, the slimy mass will pull out in the form of a glistening, mucus-like, brown thread or 'rope' (Plate III). The rope may be light or medium brown, according to the length of time which has elapsed since the death of the larva. The ropy condition is succeeded by a tacky stage, as the larval remains in the cell gradually dry up, and the colour changes to dark brown. Further drying leads to the final stage, which is a very dark brown, rather rough scale lying on the lower side of the cell, and extending from just behind the mouth of the cell right back to the base (see figure). In combs left hanging in the hive for any length of time following the death of a colony from American foul brood the originally dark brown scales often develop a coating of a greyish-white mould which makes their position and shape more readily visible when the cells are examined.

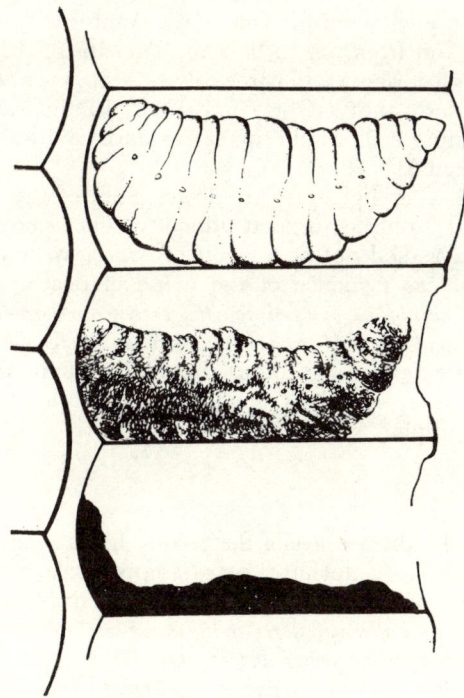

AMERICAN FOUL BROOD

Diagrammatic section through three cells of brood comb showing fully-grown healthy larva (upper cell), infected larva at ropy stage (middle cell), and scale (lower cell).

(Note the domed capping enclosing the healthy larva in the upper cell, the sunken capping of the diseased larva in the middle cell and the perforated capping of the cell containing the scale.)

The brood combs of a colony in which the disease has become established often have a patchy appearance. This is due to the presence of dead larvae in the brood nest showing all the various stages of collapse, from the ropy condition in cells with dark, sunken, or perforated cappings, to the dry scales lying in open cells whose cappings have been chewed away completely by the bees. Unsealed brood and sealed cells with normal cappings (containing brood which has escaped infection) continue to be seen for some time, but as the queen does not lay in the cells containing scales, these patches of brood gradually become constricted to small groups of cells scattered irregularly over the face of the comb.

The scales cannot be distinguished properly by a casual glance at the comb at right angles to its surface, since the cells slope slightly downwards towards the midrib, and their lower sides, on which the scales invariably lie, are partially hidden from view. They are, however, quite easy to see if the comb is held as illustrated in the upper photograph of Plate IV. Here the beekeeper is standing facing the light and is holding the frame by the lugs and side bars. He is looking down at the comb from above and the light which falls into the mouths of the cells is reflected back into his eyes. In this way the whole of the face of the comb can be scanned rapidly, from side to side and from top to bottom. The tips of the scales catch the light on their rough surfaces and can easily be seen, even when their colour is almost matched by that of the comb itself. In the old comb shown in the illustration empty cells would appear as dark cavities from which little or no light is reflected.

The scales stick very tightly to the lower sides of the cells and can be removed by the bees only with great difficulty. They persist in combs from colonies which have died out as a result of American foul brood and are, therefore, valuable as evidence of the cause of death. *Whenever a colony dies out the combs should be scanned for the presence of scales, and if anything resembling them is found a sample comb should be submitted for laboratory examination* (see page 32). The remaining combs should be left in the hive, and the latter made absolutely bee-tight pending the receipt of a report on the condition of the sample.

Spread

The spread of the disease *within* the colony has already been described. The activities of the bees during their attempts to clean out the cells containing the remains of the dead larvae lead to the distribution of the spores throughout the hive. *The honey stored in the brood combs or in the supers inevitably becomes contaminated with the spores and is, therefore, a source of infection for any colony whose bees gain access to it. This fact is of great significance in considering the spread of the disease from one hive to another, and from one apiary to another.*

If the beekeeper fails to notice the presence of the disease, or if he neglects to take steps to deal with it, the colony will become so weak that it will be unable to defend itself should it be attacked by robber bees from strong colonies nearby, either in the same apiary or in that of a neighbouring beekeeper. The robbers will take back with them all the honey they can find and some of it will be used in the preparation of brood food for feeding their own larvae. These larvae, fed on the brood food contaminated with the spores from the stolen honey, will become infected; and from this point the whole process of a decline in strength, followed by robbing and further spread of the disease, may be repeated. Thus it comes about that a strong colony—one capable of robbing a weak one—may be the first to bring American foul brood into an apiary from some source of infection elsewhere.

The spread of the disease may also be caused by the exposure of contaminated honey by the beekeeper. If honey in the super combs of an infected hive is extracted, for example, the extractor becomes contaminated, and bees which find their way into the extracting shed may carry the spores with them back to their own hives. The wet super combs after extraction will also be a source of infection for any colony to which they are given for cleaning up before storage for the winter. For this reason it is wise to label all supers with the number of the hive from which they came, and to put them back on the same hive for cleaning. Then, if any disease is discovered during the examination of the brood combs before the autumn feeding, the supers from that hive can be located and dealt with accordingly.

Failure by the beekeeper to recognize the disease when 'the first signs of trouble appear may lead to its spread in other ways. He may transfer combs of brood, or of honey, to other colonies which require strengthening and by this means spread the disease directly throughout his apiary. His appliances may become contaminated, too, and act as intermediate agents carrying spores in smears of honey and propolis.

Bees issuing in a swarm from an infected colony—a strong one, for instance, which has picked up the disease by robbing—will take contaminated honey with them, which may infect the brood produced after the swarm is hived, especially if drawn combs are used in which the queen can lay eggs without delay. Young bees coming out for a play-flight after being engaged in cell-cleaning operations in the hive, and drifting back into other hives nearby, are also potential agents for the spread of the disease.

American foul brood, like other bee diseases, may also be spread from one locality to another by the purchase of bees from an infected source, or by the sale of secondhand combs and equipment. When buying bees, therefore, care should be taken to deal with a reputable breeder; or a personal inspection of the colonies, preferably in the presence of an experienced beekeeper, should be made before making the purchase.

Treatment

A colony of bees in which American foul brood is found to be present should be destroyed by fire as soon as possible. All the contents of the hive—bees, combs quilts and honey—should be burned, and the hive parts disinfected before they are used again (see pages 7 and 8).

No less drastic way of dealing with an outbreak has been found which can be generally recommended as a completely safe and reliable alternative. It is very important, if unnecessary losses are to be avoided, that the disease should be detected in the early stages of its attack on the brood nest, before it can spread to other colonies in the apiary or to those belonging to neighbouring beekeepers. Successful limitation of the spread of American foul brood depends, in fact upon early diagnosis followed by prompt action to dispose of the disease brood and all other possible sources of infection within the hive.

EUROPEAN FOUL BROOD

Cause

European foul brood is the result of a bacterial infection of the young larvae which kills them when they are about 4 days old. The bacteria present in the dead larvae show a variety of forms, and the picture presented under the microscope, as well as the symptoms seen in the comb, may vary between one larva and another, or between one case of the disease and another. The characteristics of the disease are, therefore, not so uniform and regular as those of American foul brood.

In the early stages of infection the most conspicuous organism to be seen under the microscope is *Streptococcus pluton*, together with variable numbers of *Bacterium eurydice*. The disease is caused by *Streptococcus pluton* and the death of the larvae is probably accelerated by *Bacterium eurydice* and often by other bacteria. Later, the spore-forming bacterium *Bacillus alvei* is frequently found as a secondary invader of the tissues of the dead larvae.

Symptoms

Shortly before death the infected larva becomes restless and wanders about inside its cell instead of remaining in the normal coiled position characteristic of a healthy larva of the same age. Consequently, when it dies, about 4 days after hatching from the egg, it is found in an unnatural attitude—across the mouth of the cell, twisted spirally around the walls, or stretched out lengthwise from the mouth to the base (Plate IV). The plump, pearly-white appearance is lost; the larva collapses as though it had been melted, it turns yellowish-brown and eventually dries up to form a loosely attached brown scale. In the scale the tracheae of the larva can sometimes be seen as a network of glistening threads. The consistency of the recently dead larva varies; it may be sticky or porridge-like, but not ropy. The smell of the larval remains also varies, both in intensity and in quality; it may be exceedingly foul or merely sour, depending upon the type of bacterium which has become established after the initial infection.

The disease is essentially one of the *unsealed* brood, but some of the infected larvae may survive until after their cells are sealed, in which case there may be a few sunken or perforated cappings to be seen.

Spread

Within the hive the disease may spread to such an extent that there is a serious reduction in the number of workers emerging from the combs, followed by the decline and death of the colony. In some cases, after an appearance of the disease in the spring or early summer, the symptoms may gradually disappear during the course of the season, until no visible signs are left by the autumn. A reappearance is, however, likely in the following spring, and in the meantime the disease may have spread to other colonies in the apiary, either in an active, visible form, or in a 'dormant' form which may become active at a later date.

The factors already considered in connection with the spread of American foul brood usually apply also to the spread of European foul brood. Two points of difference should, however, be mentioned. First, the scales of European foul brood are easily removed by the bees and, therefore do not persist as visible evidence of the disease, although the combs may still be capable of carrying it. Secondly, many infected larvae pupate and develop into workers, although they leave the cell in which they were reared contaminated with bacteria. These bacteria are spread to other new larvae. This build-up of infection occurs especially in early spring. The rapid spread of disease which often seems to occur in apiaries when the disease has once made its appearance is because many larvae in contact colonies that appear normal are in fact infected, and they all begin to die about the same time for reasons linked with the rapid increase of brood rearing in late spring.

Treatment

Colonies with a high proportion of diseased brood are best destroyed, as with American foul brood, but more lightly diseased colonies may be treated with an antibiotic under the terms of the Foul Brood Disease of Bees Order, 1967. Treatment is carried out by an Appointed Officer under the Order, using drugs officially dispensed following confirmation of European foul brood in sample combs submitted for diagnosis to an approved laboratory. The repeated use of antibiotics annually as a preventative treatment is, however, undesirable on the grounds that this might result in the development of resistant strains of the bacteria that cause the disease.

Destruction of Stocks Infected with Foul Brood

Destruction must be undertaken with great care, in order to prevent the spread of disease to healthy stocks by robbing, or by the use of equipment which has not been thoroughly disinfected.

Killing Bees

In the daytime, close the entrance of the hive to about 2 in. and place a piece of perforated zinc over the feedhole in the quilt or crownboard. The latter should fit perfectly so that no bees can escape into the space under the roof. In the evening, when all the bees have returned, block the entrance securely with a clod of earth. Remove the roof and packing, and pour about a quarter of a pint of petrol through the holes in the zinc over the feedhole. Replace the packing and roof so that no fumes can escape. In a few minutes all the bees will be dead.

Make certain that the hive is bee-tight and leave it undisturbed until the combs and dead bees can be dealt with. If the hive is in a bad condition and cannot be made bee-tight, it should be removed bodily and kept under cover in a place where it will be safe from the attention of robber bees.

Burning

Dig a hole in the ground about a yard in diameter and 18 in. deep. Lay straw or paper in the centre and arrange the combs on end around it. Put all the dead bees, quilts, etc., on top, and if the hive or brood chambers are old and decrepit, place them on the pile as well, together with any super combs which may have been used on the infected colony. Set the pile alight, and when everything is completely reduced to ashes fill in the hole immediately with soil.

It is important that no honey, wax, or dead bees should be dropped on the ground when transferring combs from the hive to the burning-pit, and that the burning should be done when the bees are not flying—if they are active they will be attracted by the smell of honey and wax, and may rob the pile of contaminated material.

No attempt should be made to disinfect diseased combs, since all methods so far suggested have been found to be unreliable. When handling infected stocks or material, the beekeeper should avoid soiling the clothes with contaminated honey or propolis, and should always wash the hands, hive-tool, smoker, etc., when the work is finished.

Disinfection of Hive and Appliances

The hive should be treated as soon as possible after the destruction of the bees and combs. Any brace comb or propolis should be scraped off with the hive-tool, the scrapings being carefully collected and burned. Next, the interior surfaces of the hive, including floor, brood chamber, supers and division boards, should be scorched with a painter's blowlamp. The wood should be thoroughly scorched, but not unduly blackened by charring, special care being taken to see that all cracks and corners are reached by the flame.

Appliances which cannot be treated with the blowlamp, such as the queen-excluder, feeder, smoker, and hive-tool, should be scrubbed thoroughly with a strong brush dipped in a solution made in the following way;

> Dissolve 1 lb washing soda in 1 gal hot water; add ½ lb chloride of lime, stir and allow sediment to settle; pour off the clear liquid into another vessel and apply with the brush. The solution must be freshly made and used while still hot. Care should be taken not to spill it on the clothes or to immerse the hands in it longer than necessary as it is rather caustic and has a powerful solvent action.

After the appliances have been scrubbed clean, they should be thoroughly washed in clean water before drying. As the solution acts rapidly on aluminium, appliances made of this metal should be rinsed *immediately* after scrubbing.

Precautionary Measures Against Spread of Foul Brood

Keep the apiary clean and tidy. Never throw propolis or brace comb on the ground, where it may be robbed: place it in a suitable container and remove it from the apiary.

Never buy old combs.

Never buy colonies of bees unless it is known that they come from disease-free apiaries, and never accept stray swarms of unknown origin.

Always disinfect secondhand hives before use, by the method described above.

Never feed honey from doubtful sources, or allow bees to gain access to it.

If a colony dies during the winter and the trouble is not due to starvation, close the hive, pending the examination of a sample comb, to prevent the remaining stores being robbed out.

Never exchange brood or super combs between one colony and another unless it is known that all colonies in the apiary are free from disease. Where possible, supers should be marked and always used on the same colonies.

Care should be taken to prevent robbing at all times.

The hives should be arranged in such a way that drifting is reduced to the minimum.

Always keep a careful watch on the brood for signs of disease. If there is the least suspicion that all is not well, send a comb of brood to the Agricultural Development and Advisory Service, Rothamsted Lodge, Hatching Green, Harpenden, Hertfordshire, or to the Beekeeping Adviser, Agricultural Development and Advisory Service, Trawscoed, Aberystwyth, Cardiganshire (page 32).

Both American and European foul brood are subject to the provisions of the Foul Brood Disease of Bees Order, 1967.

FOUL BROOD DISEASE OF BEES ORDER, 1967

This Order empowers the Minister of Agriculture, Fisheries and Food to arrange apiary inspections for preventing the spread of foul brood diseases, to take samples of comb for laboratory examination where disease is suspected and, where disease is confirmed, to require destruction of the diseased colonies. In cases of European foul brood disease, treatment with an antibiotic, by the Minister's officer, of the infected bees and a subsequent treatment of other bees on the premises, may be allowed as an alternative to destruction.

Bee Diseases Insurance, Limited. An insurance scheme, designed to compensate subscribers for losses incurred through the destruction of bees and combs as a result of an outbreak of foul brood, is operated by Bee Diseases Insurance, Limited, a specialist body of the British Beekeepers' Association. Particulars of the scheme may be obtained from officers of local beekeepers' associations.

CHALK BROOD

Chalk brood takes its name from the chalky-white appearance of the dead brood. It is caused by the fungus *Ascophaera apis* (formerly known as *Pericystis apis*). It is not uncommon, but as a rule only small patches of brood are infected and it seldom causes serious losses to the beekeeper.

Larvae ingest spores of the fungus with their food. The spores germinate in the gut and the subsequent growth of the fungus causes the death of the brood after it has been capped. The dead larvae are at first rather fluffy and swollen, taking on the hexagonal shape of the cell, but later they shrink and become quite hard. They may remain chalky-white in colour, but sometimes, when two strains of the fungus grow in the same larva, fruiting bodies containing spores are formed on the surface and the colour then becomes dark grey or almost black (Plate V).

Chalk brood is readily recognized in cells containing diseased larvae after cappings have been removed by the bees. The disease usually affects only a few cells or worker or drone brood here and there on some of the combs, but occaionally extensive patches of dead brood may be found throughout the brood nest. The fungus grows only in larvae that have been slightly chilled, so that any manipulations of colonies leading to the chilling of the brood, e.g., spreading or splitting the brood-nest in spring, are to be avoided.

This precaution effectively prevents disease spreading. The disease is then transient in character, clearing up during the course of the season. Badly infected combs are best destroyed, however, and replaced by clean drawn comb or foundation. Should the disease persist to an extent that badly affects the strength of the colony, then re-queening with an unrelated queen may be tried in the hope that her progeny will prove to be more resistant to the infection than the progeny of the previous queen.

STONE BROOD

Stone brood disease is caused by a mould belonging to the genus *Aspergillus*. The mould attacks the brood, transforming the larvae into hard stone-coloured objects which are found lying in open cells. Adult bees may also be attacked and killed by the mould. The disease is rare in this country, only three cases having been recorded over the past thirty years. A sample comb from any suspected case of the disease should be sent for examination without delay (see page 32). No form of treatment is known and an infected colony should be destroyed by fire.

Since some of the *Aspergillus* moulds can cause trouble in the respiratory passages in man, care should be taken not to inhale any of the dust (spores) with which the dead brood may be covered.

SACBROOD

Studies made by the Bee Department of Rothamsted Experimental Station on sample combs selected from those sent for examination to the Agricultural Development and Advisory Service have shown that the virus disease of honeybee larvae known as sacbrood occurs in Britain, as it does in other countries in Europe and elsewhere. It seems clear that many of the brood disorders diagnosed in the past as addled brood were in fact cases of sacbrood, and that this virus infection, rarely recorded as such in the past, must now be regarded as not uncommon in this country.

Symptoms

Larvae infected with sacbrood die after they are sealed in their cells, usually in the propupal stage. They become light yellow in colour, with tough skins. The colour then darkens and the outer skin becomes very loose, forming the 'sac' which encloses a watery fluid. Gradually the larvae dry and shrivel to form thin scales resting on the lower angle of the cells but with the head ends turned upwards. At this stage they are often visible through holes in the capping or in open cells from which the cappings have been completely removed by the bees. The complete scale resembles a miniature 'Chinese slipper'; it is readily removable intact from the cell with a pair of forceps.

The infection is seldom apparent at the beginning of the season. It may be noticed from May onwards, when the post-mortem changes leading to the formation of scales become visible in cells with perforated or torn-down cappings. The scale is not infective, in contrast to those of American and European foul brood. The virus is not able to survive for more than about three weeks in honey.

Treatment

No specific form of treatment is known for sacbrood. Only a small proportion of the total amount of brood is affected as a rule, and the infection usually disappears towards the end of the season. However, if the infection is persistent, or if extensive areas of dead brood are found throughout the brood-nest, it is recommended that the colony be re-queened as soon as possible with a queen from a healthy source.

ADDLED BROOD

Addled brood is a disorder which at first sight may be mistaken for American foul brood, since it is the sealed brood which is affected, and moist, sunken, or perforated cappings are usually to be seen. The dead brood maintains its shape and outline to some extent, however, and the contents of the affected cells are never ropy.

Cause

The condition is the result of some defect in the egg which does not become apparent until the larva has completed its development up to the propupal or pupal stage. (Occasionally cases are reported in which the egg never matures and fails to hatch, but whether this phenomenon is an extreme form of the addled brood described here is not known.) The queen which lays the eggs is, therefore, to be regarded as the cause of the trouble.

Symptoms

Both worker and drone brood may be affected. The brood normally dies in the propupal or pupal stage, that is, after the cells have been sealed. Combs of sealed brood taken from diseased colonies often have a 'pepperbox' appearance due to the patchy arrangement of the sealed cells. The dead propupae are soft and moist, and are difficult to remove from their cells without damage. One of the outstanding features of the disease is that advanced pupae and bees apparently almost ready to emerge are found dead in their cells. In such cases the pupae are often small with the abdominal parts much reduced in relation to the size of the head and thorax. The bees usually attempt to remove the dead brood by pulling off the cappings and chewing the cell contents.

Treatment

The queen should be destroyed and replaced by one which produces normal brood—preferably one reared in another colony, since daughter queens reared in the affected colony may lay eggs carrying the same defect as those laid by the parent queen.

As the disorder is not contagious and as the bees will gradually remove the dead brood from the cells, there is no need to destroy any of the combs, or to take any special precautions when manipulating affected colonies. The appearance of addled brood in several colonies in the same apiary would suggest that the queens are derived from a common stock, and that requeening with a different strain of bee would be advisable.

Comparative summary of the chief symptoms of four brood diseases commonly occurring in Great Britain

	American Foul Brood	European Foul Brood	Chalk Brood	Sacbrood
Time of death	In propupal stage or shortly after pupation—*after* cell has been sealed	Usually about 4 days after hatching from egg; essentially disease of *unsealed* brood	After sealing	Propupal stage
Cappings	Sunken, dark in colour, often moist and perforated	Usually none visibly affected; if larvae die after sealing—cappings, dark, sunken, and perforated as in American foul brood	Not affected, but later removed by the bees	Perforated or removed completely
Position of dead brood in cell	Always in lower angle; stretched out along length of cell	May occupy any position; often twisted spirally; collapsed or melted appearance	Normal	Normal
Colour changes	From light creamy-brown to dark coffee-coloured; eventually almost black	Variable; yellow-brown to brown or dark brown	Yellowish-white at first (before removal of capping), then chalky-white; may become greenish-brown or black	Brown, darkening to almost black
Consistency	Slimy; marked ropiness at coffee-coloured stage, then becoming tacky	Soft, sticky or porridge-like; no ropiness	Spongy or rubber-like mummies; later tough and fibrous and easily removed from cells	Watery fluid within loose skin
Scale	Hard, dark brown and adherent; always in lower angle of cell	Brown; position variable; tracheae often visible; easily removed	—	Brown to almost black; head turned upwards, abdomen flattened; easily removed

Abnormal Conditions of the Brood not caused by Infectious Disease

CHILLED BROOD

CHILLED brood may be caused in a number of ways and is very often the result of carelessness or inexperience on the part of the beekeeper. For example, hives may be opened too early in the spring, and the combs left exposed to a cold wind during the examination; or the brood may be 'spread' prematurely by the insertion of empty combs into the brood nest in the hope of stimulating the queen to increase her egg-laying.

Chilling may also occur if there is a serious loss of bees caused by disease, or by poisoning, when the remaining bees may be too few to cover the brood. A similar lack of balance between the amount of brood and the number of bees may occur if a prolonged spell of warm weather in the spring is followed by unseasonable weather later; the bees raise more brood than they can cover adequately in the event of a sudden cold period, the arrival of which causes them to contract towards the centre of the brood nest and to leave some of the cells unprotected.

Young coiled larvae which have been chilled often turn a glistening black and are difficult to see unless the brood combs are new; in old combs their colour matches that of the dark interior of the cells. Older chilled larvae become greyish in colour. The sealed brood may show moist, sunken and perforated cappings, as in American foul brood, but the cell contents are never ropy.

The trouble will usually be cleared up by the bees themselves, but in severe cases the affected combs should be replaced by clean ones, and either rendered down for wax or kept in a dry place for use again later when the dead brood has become shrivelled and easy for the bees to remove.

STARVED BROOD

Colonies which have produced a large amount of brood in the spring may find themselves short of food during bad weather in the early summer. Some of the brood may then have to be abandoned: egg-laying ceases, young larvae are often eaten by the bees and pupae may be found thrown out on the ground in front of the hive. Starved or neglected brood may also result from such operations as artificial swarming and certain systems of swarm control, if they are carried out without due care.

It should be realized that in our variable climate bees sometimes run short of stores in May or early June, or even later, and a supply of sugar should be kept available for feeding in such an emergency.

NEGLECTED DRONE BROOD

The presence of a drone-laying queen or of laying workers leads to the production of irregular patches of drone brood. If the colony is weak, many of the larvae and pupae reared in worker cells are small and undernourished. If this state of affairs is allowed to continue, the brood becomes neglected by the bees and chilling causes death and subsequent decomposition. The combs have an untidy appearance due to scattered groups of cells with raised and irregularly dome-shaped cappings, many of which may be partially torn away leaving the heads of the dead drones exposed to view. The decomposing larvae become soft and often brown, and eventually they dry up to form dark masses sticking to the sides of the cells.

Sometimes there are no obvious signs of drone pupae among the sealed brood; all the cappings are uniform in shape with nothing to indicate that the pupae underneath are not all workers. Dead brood is present in open cells, with an appearance similar to that of European foul brood, but microscopic examination shows none of the bacteria characteristic of this disease. If, however, the sealed cells are uncapped, a few drone pupae are found in what at first sight seems to be a patch of sealed worker brood. This condition indicates that the queen is defective in that she is laying drone eggs here and there in worker cells. Some of these eggs hatch and develop into small drone pupae; others fail to develop properly, the larvae dying in open cells before they reach the pupal stage.

The obvious remedy is to requeen or to unite the remaining bees with another colony, but the inexperienced beekeeper may overlook the real cause of the trouble—the lack of a good fertile queen—and recognize only the more obvious decomposition of the brood, with its superficial resemblance to foul brood. Badly distorted combs and those containing a large amount of dead brood are best replaced by clean drawn comb.

BALD BROOD

In the normal course of events, the honeybee larva reaches its full size 5 days after hatching from the egg and is then capped over by the workers. The change from larva to adult bee then occurs and in about a fortnight the fully developed bee emerges and takes its place amongst its fellows. Between capping and emergence, the brood is normally hidden from view. Sometimes, however, it happens that the heads of the developing pupae and even those of the pre-pupae are visible (Plate VI). This condition is known as bald brood and occurs in one of two ways.

Wax moth grubs, burrowing in the midrib of the comb and in the cell walls, have a disturbing effect on the development of the adult bee. Bees about to emerge are found in unsealed cells, often with deformed wings and legs; faecal pellets, deposited by the wax moth grubs, may be seen sticking to their bodies (Plate VI). The cappings of the cells are probably consumed by the wax moth grubs. The bald pupae occur in patches or lines, depending on the route taken by the grubs in burrowing through the combs.

Occasionally a colony is found in which the bees do not cap the larvae normally in the first place. Sometimes the edges of the cells are turned slightly inwards, sometimes the cappings are complete except for a small hole in the centre. Some of the brood may die, but most of it will develop quite normally. Requeening with a mated queen of a different strain will remove this undesirable characteristic from the colony.

Adult Bee Diseases and Treatment

INVESTIGATIONS into the cause of disease in the adult bee date from the early years of this century. The widespread attention attracted by 'Isle of Wight Disease'—so called from the area in which heavy losses of bees were reported in 1905—greatly stimulated research work in Britain and elsewhere. Several distinct maladies, separable with the aid of the microscope but having some symptoms in common, came to be recognized. The designation 'Isle of Wight Disease' had long since ceased to be applicable and has given place to other names indicating the specific nature of the causal agents of these diseases.

More recently there have been further advances which have helped to fill some of the remaining gaps in the picture. New treatments have been developed, diagnosis and advisory services have been expanded, and in all these respects beekeepers now have greatly improved facilities available to them for the detection and contol of adult bee diseases.

ACARINE

'Acarine' is the term commonly used to denote an infestation of the adult honeybee by a parasitic mite inhabiting the breathing tubes of the thorax. Recent work has shown that, contrary to the former widely accepted belief, the mortality of infested bees during the summer months, and their foraging activity, do not differ appreciably from those of normal bees. It is mainly during the winter and early spring that the effects of heavy infestation become apparent. Colonies in which more than 30–40 per cent of the bees are infested with the mite are then less likely to survive than non-infested or lightly infested colonies.

Among the factors conductive to the development of a damaging level of infestation are poor summer seasons, when the bees spend long periods of inactivity in close contact with one another on the combs (see section on 'Spread' page 17), and restricted brood rearing due to periods of queenlessness, when the increased length of life of adult bees allows more mites to breed in them. It appears, however, that in otherwise normal colonies the infestation is usually suppressed naturally to a level at which no effect on colony performance is detectable.

The mite is widespread in colonies of bees throughout Britain. Of the 45,700 samples of bees submitted for examination to the various adult bee disease diagnosis centres in England and Wales (see Appendix page 32) during the ten years 1958–67, 3,850 (8·4 per cent) were found to have some infested bees. However, most colonies contain few infested bees and appear healthy. The methods described for the treatment of colonies (page 19) are effective in reducing the likelihood of an infestation rising to a damaging level in apiaries where the mite is present. It should be realized, however, that a heavy infestation can be the *result* of colony weakness rather than its immediate cause, and that treatment directed solely against the mites, however successful it may be in killing them, is no substititute for sound

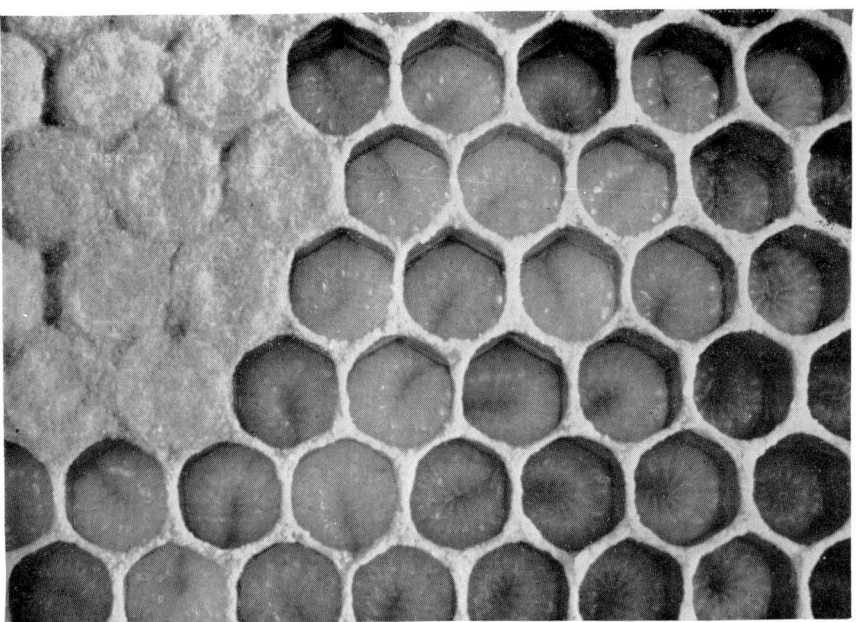

HEALTHY BROOD

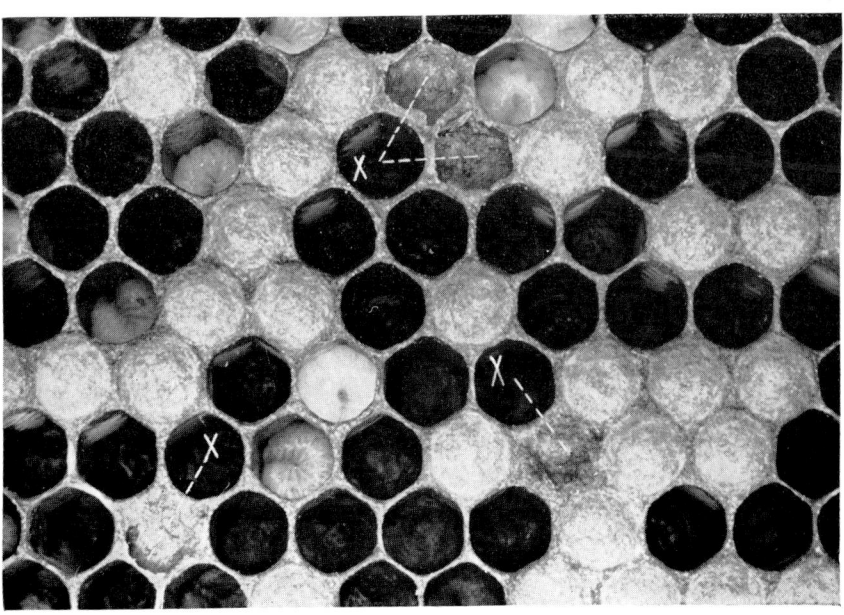

AMERICAN FOUL BROOD
Portion of comb showing four sealed cells (×) with cappings becoming dark and sunken

PLATE I

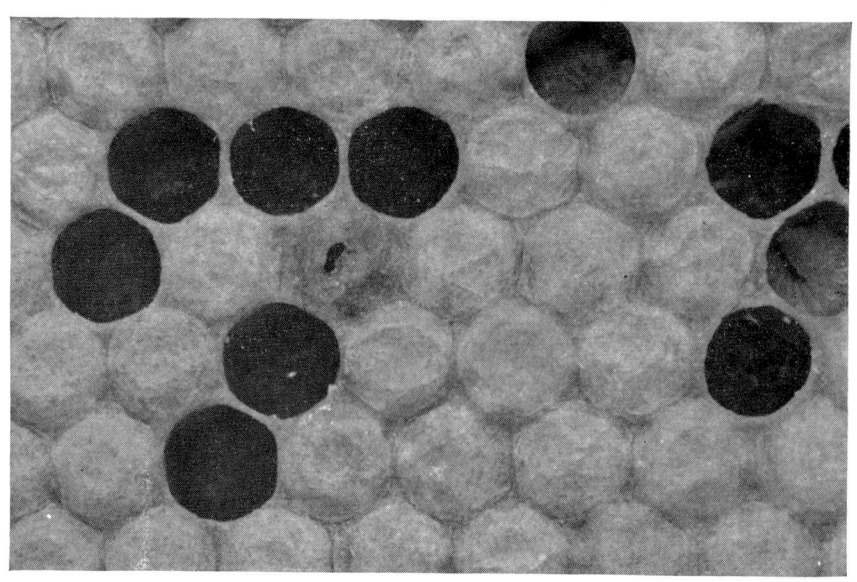

AMERICAN FOUL BROOD
Close-up of comb showing one cell with dark, slightly sunken, perforated capping

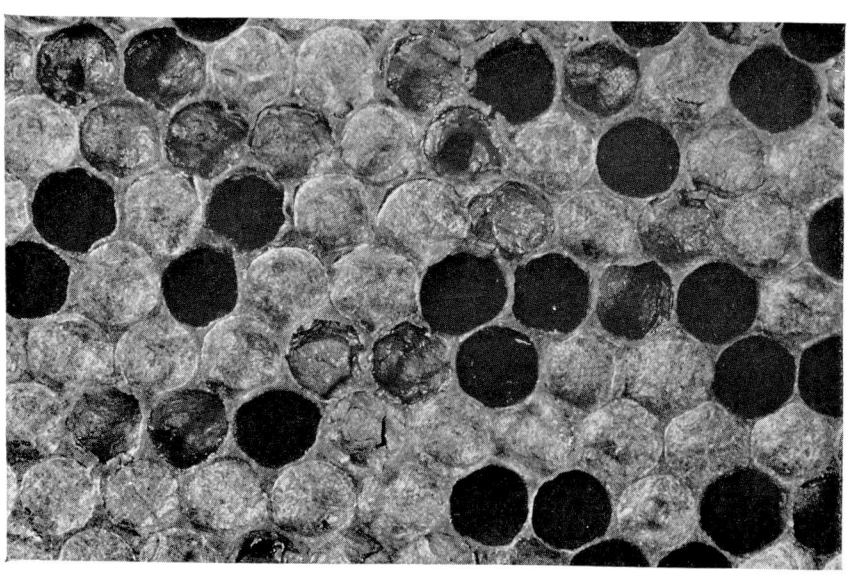

AMERICAN FOUL BROOD
Many sealed cells with dark, sunken or perforated cappings

PLATE II

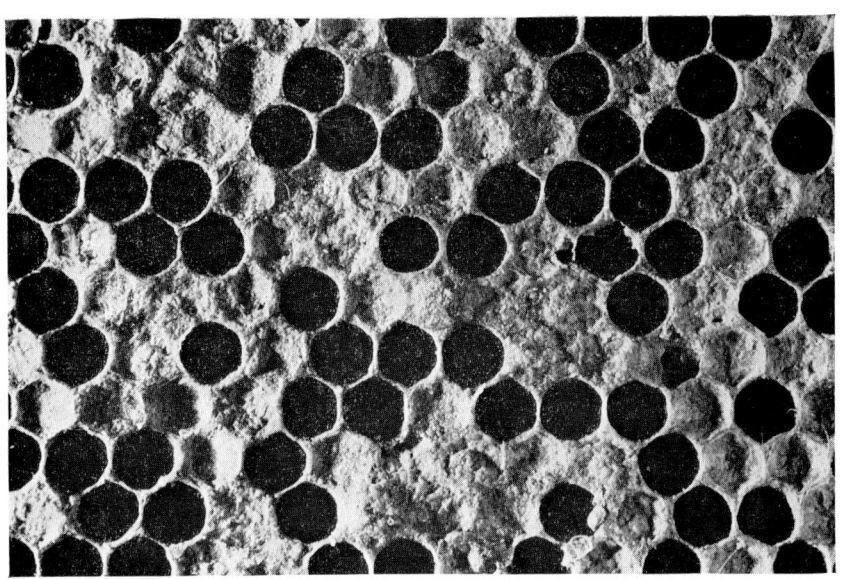

AMERICAN FOUL BROOD
Open cells with cappings removed by bees and containing scales (not visible here)

AMERICAN FOUL BROOD
The 'ropiness' test

PLATE III

AMERICAN FOUL BROOD
How to examine comb for the presence of scales (see page 4)

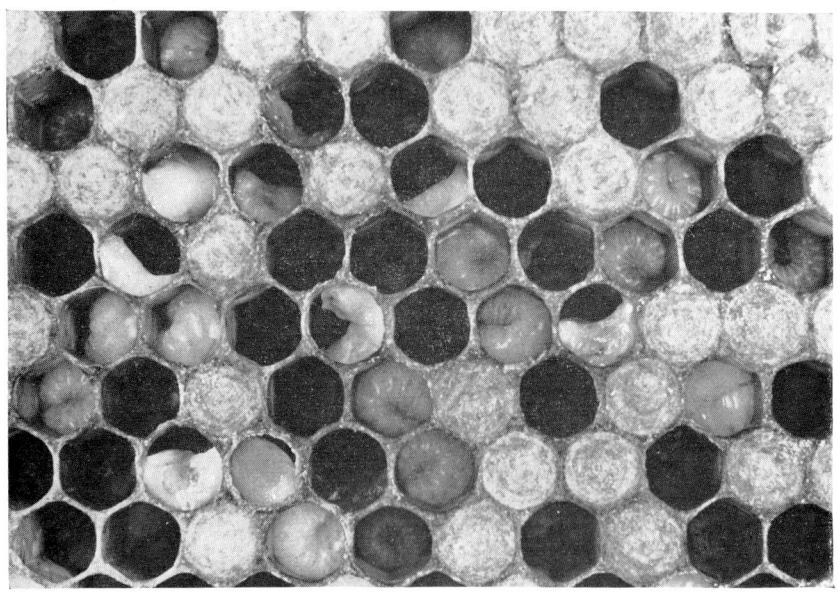

EUROPEAN FOUL BROOD
Note the unsealed larvae in various stages of collapse

PLATE IV

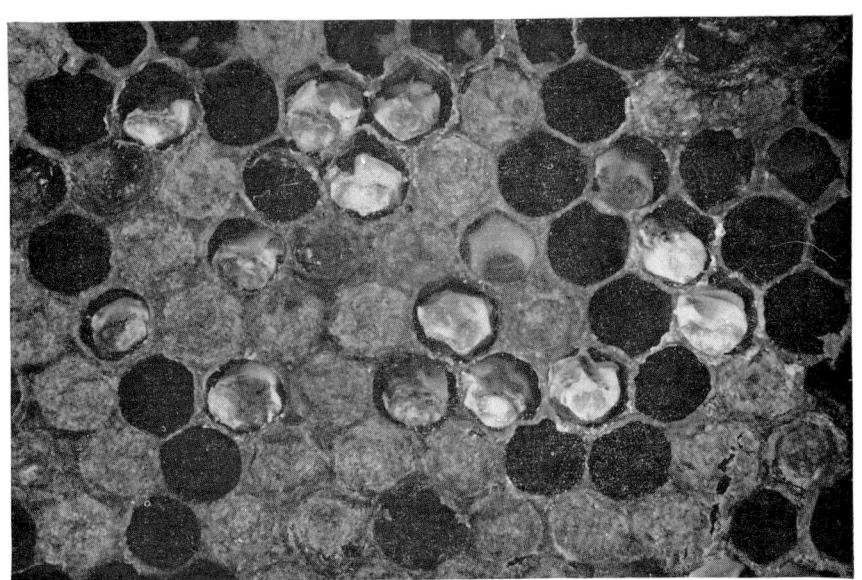

CHALK BROOD

White 'mummies' in cells uncapped by bees

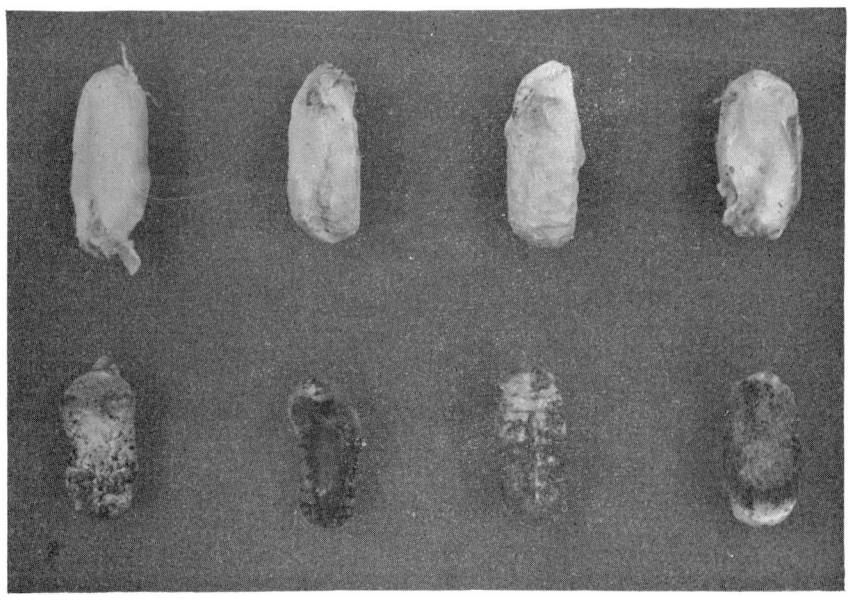

CHALK BROOD

'Mummies' removed from their cells

PLATE V

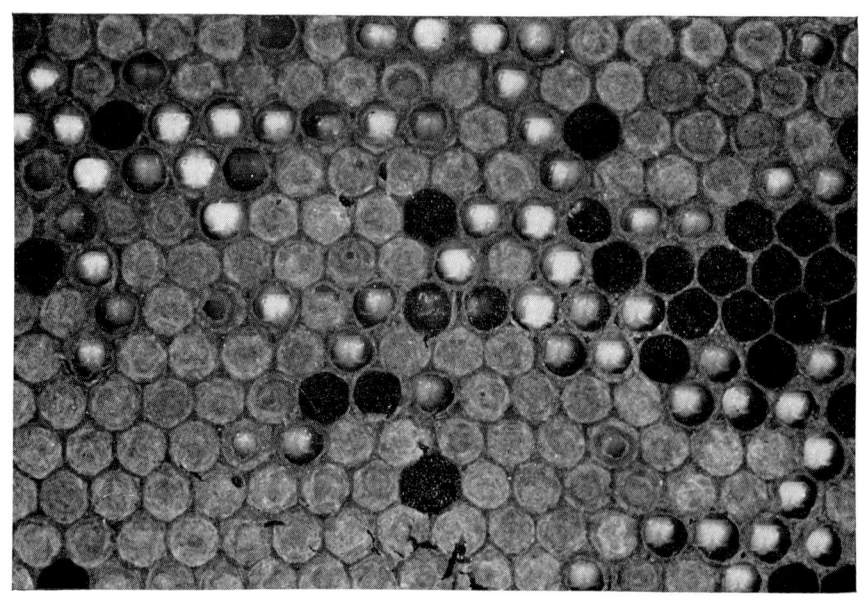

BALD BROOD

Bald brood due to the presence of wax-moth larvae

BALD BROOD

Section of comb showing wax-moth larva, holes made by it in cell walls, and faecal pellets deposited on worker pupae

PLATE VI

EXAMINATION OF BEES FOR ACARINE
Apparatus

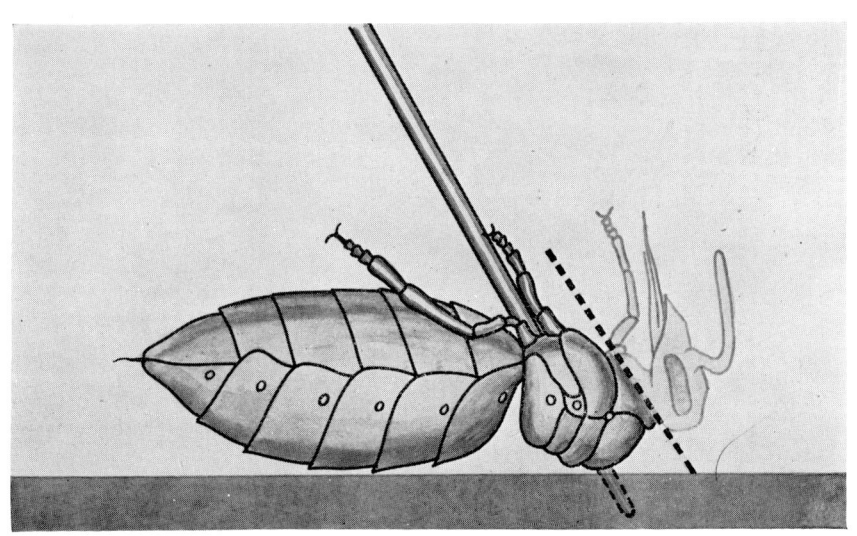

EXAMINATION OF BEES FOR ACARINE
Removal of head and first pair of legs

PLATE VII

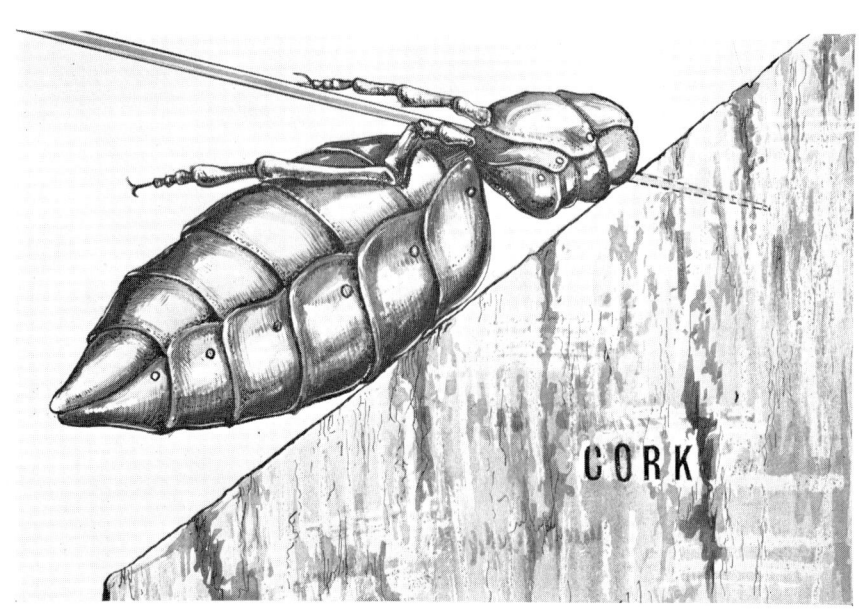

EXAMINATION OF BEES FOR ACARINE
Viewing position for dissection

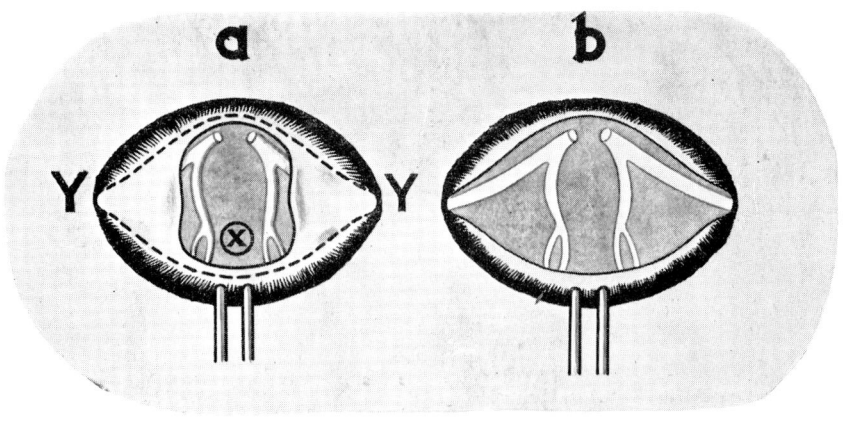

EXAMINATION OF BEES FOR ACARINE
Removal of 'collar'

PLATE VIII

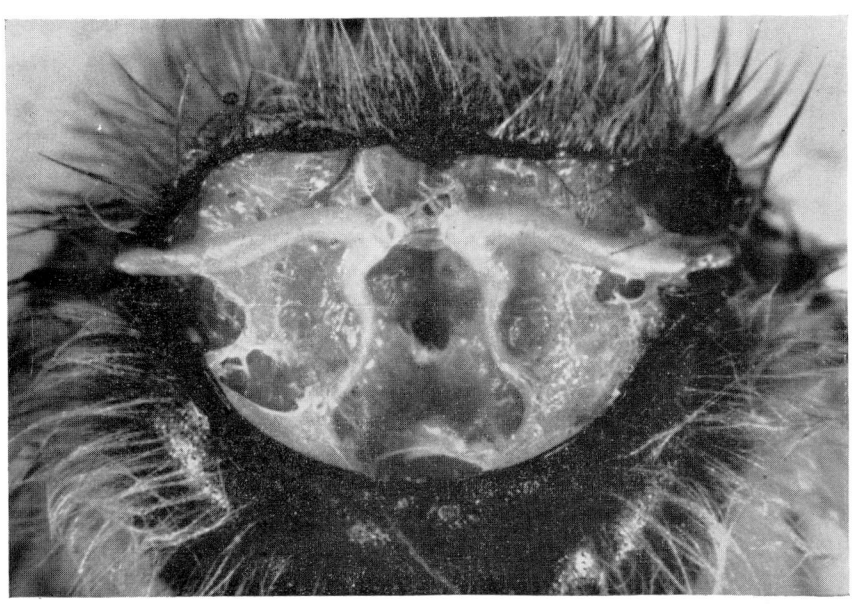

EXAMINATION OF BEES FOR ACARINE
Healthy tracheae

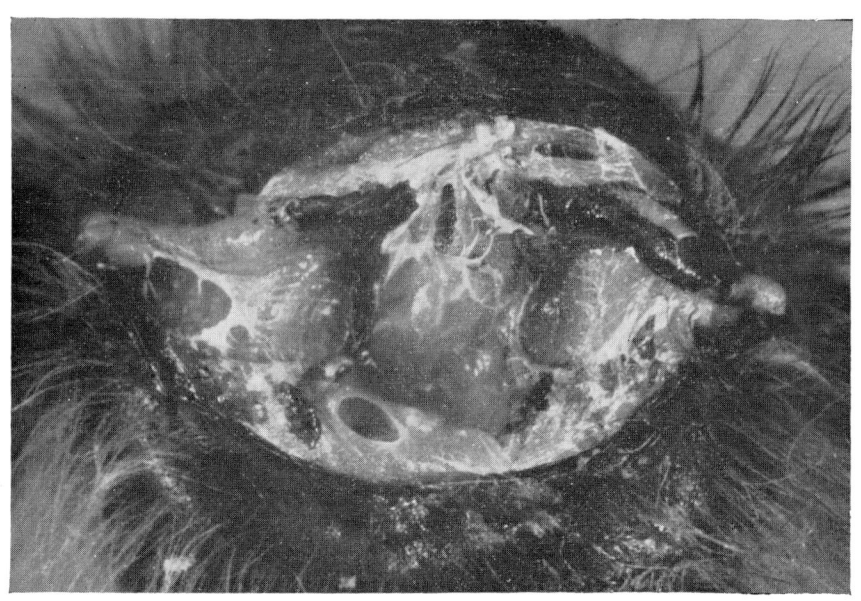

EXAMINATION OF BEES FOR ACARINE
Bilateral infestation

PLATE IX

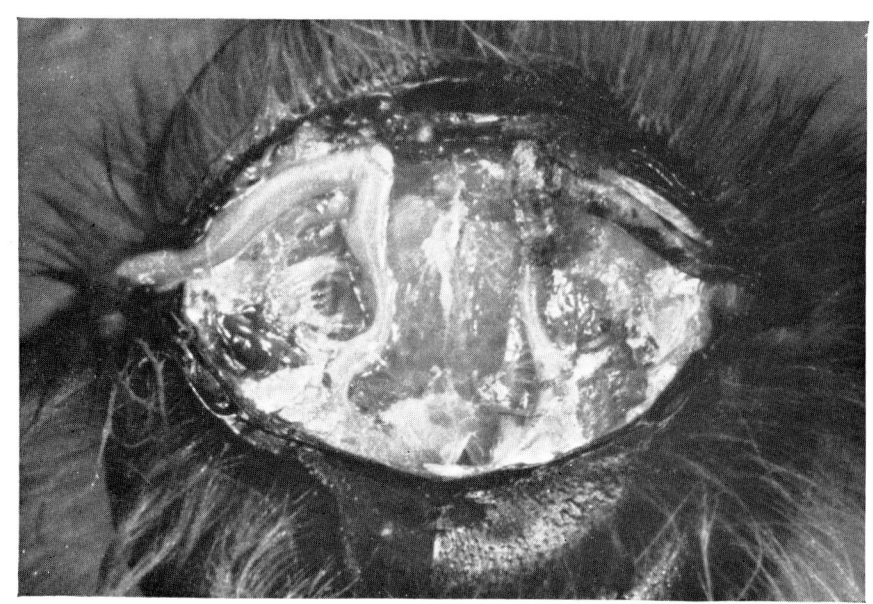

EXAMINATION OF BEES FOR ACARINE
Unilateral infestation

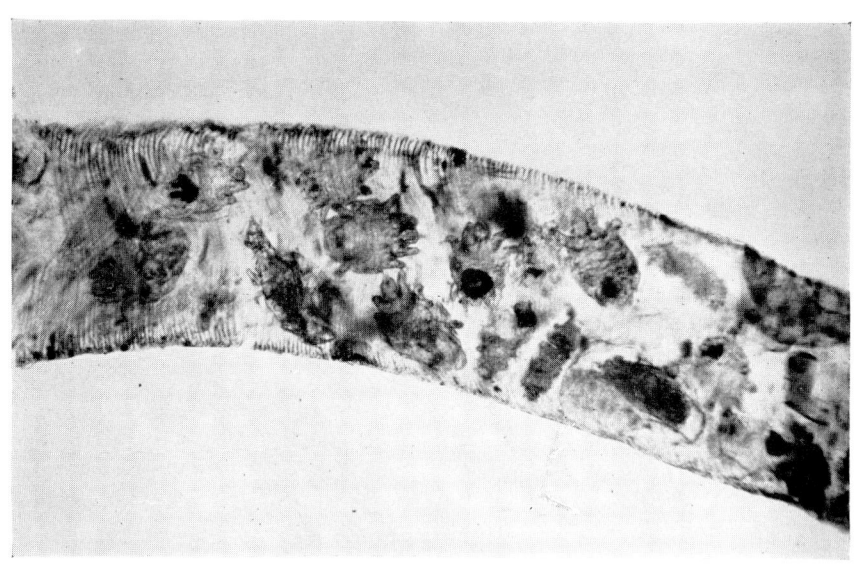

EXAMINATION OF BEES FOR ACARINE
Infested trachea (× 100)

PLATE X

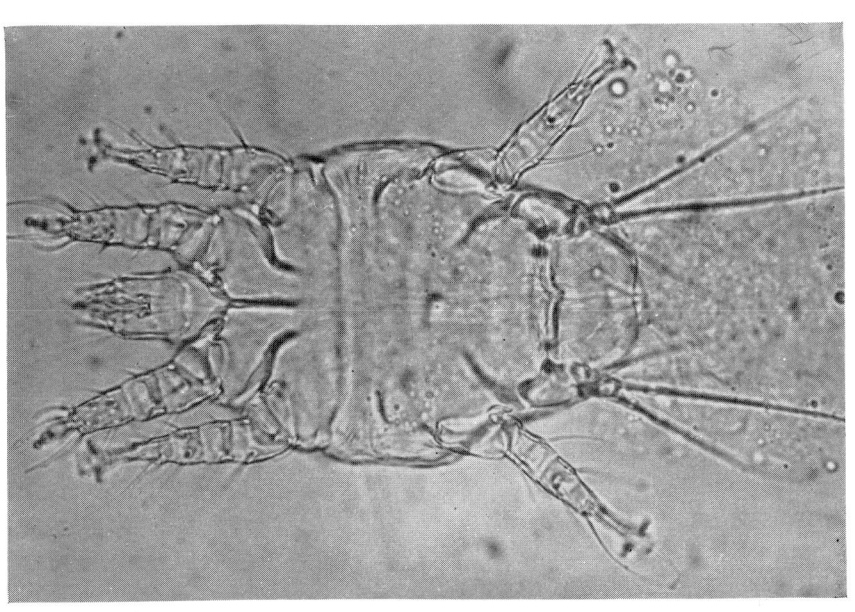

ACARINE

Acarapis woodi female (× approx. 500)

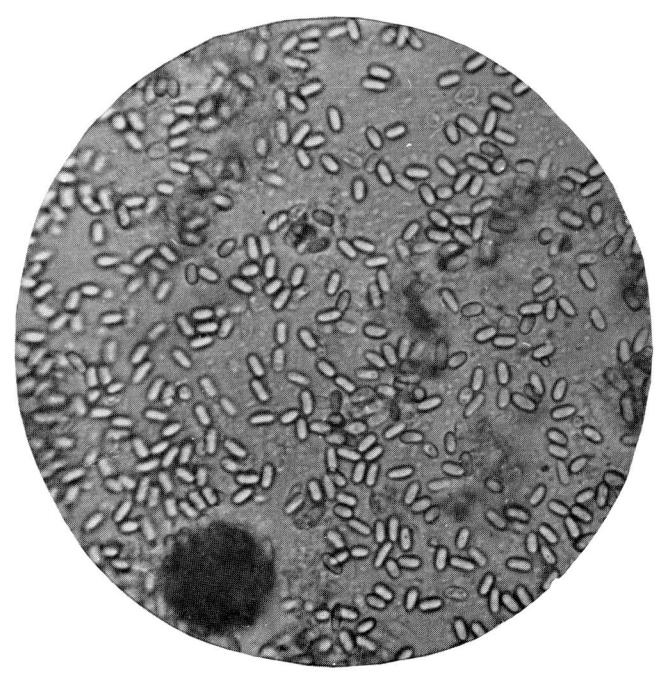

NOSEMA

Spores of *Nosema apis* (× 1000)

PLATE XI

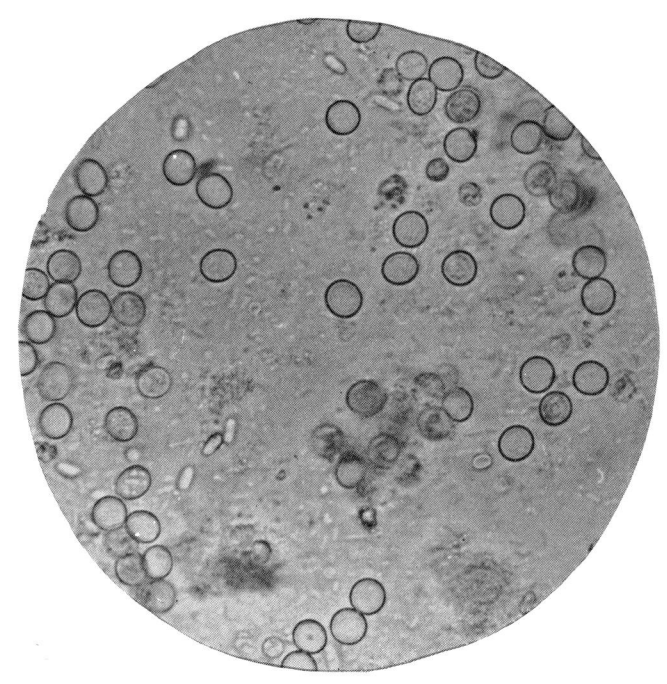

AMOEBA

Cysts of *Malpighamoeba mellificae*, with some spores of *Nosema apis* (× 1000)

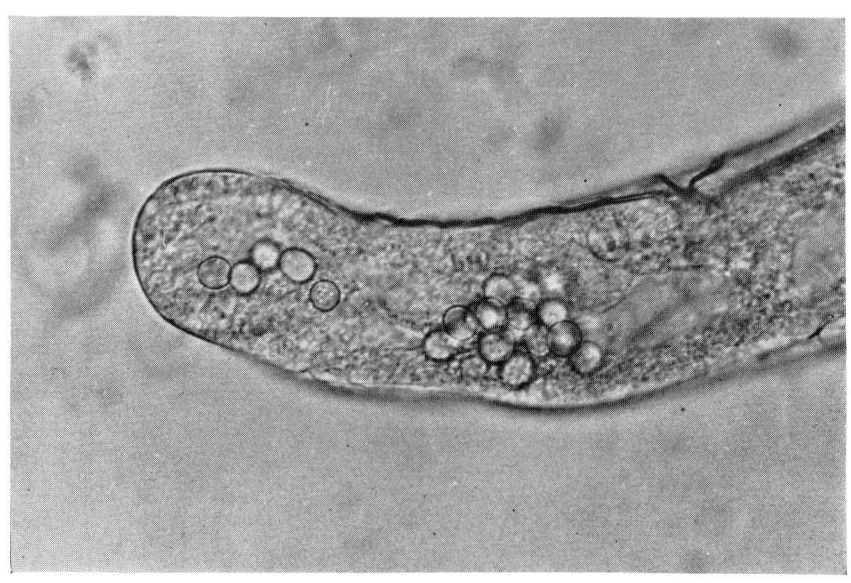

AMOEBA

Cysts of *Malpighamoeba mellificae*, in distal end of Malpighian tubule
of worker honeybee (× 1000)

PLATE XII

beekeeping methods aimed at maintaining strong colonies headed by vigorous queens.

THE MITE

The respiratory system of the adult bee consists of a complex arrangement of breathing tubes (tracheae) and air sacs. These carry air to all the organs of the body from a series of paired openings, known as spiracles, situated on the thorax and abdomen. The mite *Acarapis woodi* Rennie, breeds in the tracheae leading into the thorax from the first pair of spiracles between the first and second thoracic segments.

The adult female mite crawls through the spiracle, attracted to it by a pulsating flow of air produced by the respiratory movements of the bee, enters the trachea beyond and soon starts to lay eggs. From the eggs emerge young mites, known as nymphs, which go through a series of moults before reaching their adult form as mature males or females. The nymphs and adults feed on the blood of their host. They have suctorial mouth-parts which they apply to punctures made in the wall of the trachea by a pair of fine, retractable stylets enclosed by the mouth opening.

In an infested bee the mites are usually present in both the left and right tracheal trunks of the thorax, but an infestation in the trachea of one side only is not uncommon. Queens and drones can become infested, as well as workers.

SYMPTOMS AND DIAGNOSIS

Bees from colonies that contain many infested individuals may be found in the apiary, clinging to the stems of plants, or crawling about with fluttering wings on the grass near the hive. The abdomens may be distended, and the wings often have a dislocated appearance with the hind wing held at an abnormal angle to the body. Such 'crawlers' may emerge from the hive in large numbers in the autumn or spring on bright days with a low shade temperature.

These symptoms, however, are not characteristic of acarine, since very similar symptoms can arise from other, unrelated causes. Heavily infested bees usually appear normal, especially in conditions when bees are actively flying. The detection, under a low-power microscope or good hand-lens, of the actual presence of the mites, following a simple dissection of the bee to expose the thoracic tracheae is, therefore, the only certain way of diagnosing the infestation (see page 21).

The thoracic tracheae of a healthy bee have a uniformly smooth, creamy-white appearance, whereas those of an infested bee undergo progressive deterioration, marked at first by translucent areas indicating the position of the individual mites within, and later by an irregular, patchy discoloration of the tube as the mites increase in numbers. Eventually the tracheae may become quite black and full of mites and the by-products of their feeding activites.

SPREAD

Within the Colony

The breeding place of the mite has already been described. On reaching maturity inside the tracheae, some of the female mites crawl through the

spiracle and attach themselves to other bees in their attempts to find new
hosts in which to lay their eggs. They can succeed in this only if they find
their way on to the thorax of a bee which is less than 5–6 days old, though
the reason for this immunity of older bees is not known.

The spread of the infestation depends, therefore, upon the presence of
young bees in the hive, and on the close contact of these young bees with
old bees which are already infested. In the winter cluster, which will contain
no young bees, the mites can breed within the bees already infested in the
autumn, but there can be no spread of mites from the tracheae of one bee
to those of another.

Infestation of Young Bees

Many of the old, infested bees will die off during the winter and early
spring; but those which survive long enough will contain female mites
capable of entering the young bees which emerge following the resumption
of brood-rearing in January or February. Young bees becoming infested
then, will in turn act as sources of spread to later batches of young bees.
In this way, the infestation can be carried on from bee to bee throughout the
summer months.

During the active season, however, the average life of the workers, in
any case, is short, and the presence of the mites may not have any obvious
effect on their activities. In a good season infestation decreases spontaneously,
often very rapidly, but young bees are always available for the migration of
the female mites, especially after a poor season, the mites may well be
present in a high proportion of the bees produced in the autumn—in those
bees, in fact, on which the survival of the colony through the winter depends.

Winter bees normally have a much longer life than summer bees, but
if they are infested by the mites their life is shortened.

When the proportion of badly infested bees is high, the loss of bees is
enough to cause the collapse and death of the whole colony.

From One Colony to Another

From the time the initial infestation takes place to the time signs of
the disease are discovered there may be an interval of weeks, or months,
or, more often, no sign of disease may ever appear, depending upon the
rate at which the mites breed and spread within the colony. Consequently,
the source of the infestation can never be ascertained. Nevertheless, in all
new cases some form of contact must have occured with a previously
existing case, either in the same apiary or elsewhere.

Ways of Contact

There are several ways in which such a contact can occur. The drifting
of infested workers or drones gives the mites an opportunity of spreading
from one hive to another in the same apiary; so does the transfer of bees
from one colony to another by the beekeeper, if the mites are present in
the colony which is being depleted of bees. Again, bees from a diseased
colony may join a healthy swarm and be hived with it; or truant swarms
may carry the mites with them and act as sources of spread in a new locality.

Fortunately, the mites are not capable of living for any length of time
apart from their host bees, so that combs from colonies which have died out

from acarine are quite safe to use again later—always provided that no other disease, e.g., foul brood, or nosema, was present also.

TREATMENT

Two methods of treatment are described here. One is for use in cold weather, during the autumn or early spring; the other is for use during warm weather in early summer. Both methods involve using volatile substances as fumigants while the bees are confined to the hive.

The first hive method, the Frow treatment, has been widely used for many years. Recent work has confirmed its efficacy, but it now seems clear that the destructive effect of the fumes on the mites is followed by the death of a high proportion of those bees in the colony whose thoracic tracheae had already been damaged by the infestation.

The second method, the smoke strip treatment, is of relatively recent origin, and is based upon the properties of a chemical compound which is highly toxic to mites, yet harmless to bees when applied as recommended. The choice of treatment will depend upon the time of year at which the diagnosis of the infestation is made, but either treatment can be used as an annual routine precautionary measure.

It must be remembered, however, that even if all the mites within the infested bees are killed, the colony cannot be restored immediately to a normal condition of health. This must await the replacement of the old, damaged bees by young, healthy ones. It follows that when the percentage of badly infested bees is high, there may not be enough active and healthy bees left to enable the colony to survive. An apparent lack of success, therefore, may not necessarily be due to any failure of the treatment, but rather to a natural outcome which was inevitable before the treatment was started. Such cases can be avoided by the routine annual application of either of the methods described below, since the mites will be unable to establish themselves to any extent, and the conditions necessary for the development of a high infestation will not be allowed to arise.

Re-Queening after Treatment

If the queen in a colony with acarine is infested with the mites, the damage to her tracheae will obviously persist after treatment. Where an infested colony which has been treated and shows no further evidence of infestation fails to develop satisfactorily, it is advisable to re-queen as early in the season as possible with a vigorous young mated queen from a healthy source.

Frow Treatment

The Frow treatment is recommended for application only in the late autumn or early spring, since it is likely to induce robbing at times when the bees are active, and to disturb the cluster if used during the winter.

FROW MIXTURE

Nitrobenzene.	.	2		
Safrol .	.	1	}	parts by volume
Petrol .	.	2		

The mixture is poisonous and highly inflammable. It should be kept in a bottle labelled POISON *and handled with care. Store the bottle in a safe place out of the way of livestock and children.*

Bottles of the mixture, ready for use, can be obtained from dealers in beekeeping requisites.

Time of Treatment. Treatment should be carried out either in November or February (late October or early March in northern areas), preferably when a day on which the bees have been flying is followed by a frosty day, with a prospect of cold weather continuing for some time.

Dosage and Administration. Thirty minims (1.8 c.c.) of the Frow mixture is poured on to a flannel pad. The pad is then inverted over the feed-hole in the crown board or quilt, and covered with a tin lid or an inverted saucer. Every *other* day another dose of 30 minims (1.8 c.c.) is given on the same pad, seven doses being given in all. After the last dose has been given on the thirteenth day, the pad is left in position for a further three days and then removed.

When it is not possbile to pay repeated visits to the colonies, a *single* dose of 75 minims (4.5 c.c.) can be given. The single-dose method should be used only when the weather is cold. If used in mild weather, or in the milder parts of the country—e.g., Cornwall—there may be some risk of killing the bees in the colony as a result of too high a concentration of the vapour. The pad should be left in position for fourteen days, or up to three weeks if the weather remains cold.

A suitable graduated vessel for measuring the doses may be obtained from a chemist; or a small glass specimen tube could be marked with narrow strips of gummed paper at the 30 minim (1.8 c.c.) and 75 minim (4.5. c.c.) levels, by pouring the requisite quantities of liquid into it from a borrowed measure.

Capsules containing 75 minims (4.5. c.c.) of Frow mixture are available from Bees Diseases Insurance, Limited, to members of any affiliated bee-keepers' association, through the local secretary.

Precautions should be taken against robbing if the weather turns mild during treatment. The entrances of the treated hives should be reduced to about a double bee space, using a suitably cut piece of perforated zinc to maintain adequate ventilation of the hive. These small entrances may become blocked on the inside by dead bees falling on to the floorboards; they should, therefore, be cleared with a piece of stiff wire.

Smoke Strip Treatment

Suitable smoke strips, e.g., Folbex impregnated with the selective acaricide *chlorobenzilate*, are obtainable from bee appliance dealers (and also from Bee Diseases Insurance, Limited, by members of affiliated bee-keepers' associations).

Time of Treatment. The treatment is best applied during warm weather in early summer, before the main nectar flow, and two applications separated by an interval of one week—will give a good measure of control.

The treatment is not effective under cool conditions with inactive bees clustered tightly on the combs. The smoke should circulate quickly throughout the hive; active bees assist the circulation and inhale the smoke at the same time.

During very hot weather, or during a heavy nectar flow, the use of these smoke strips is not advisable.

The smoke has no adverse effect upon the queen or the brood, or upon stores of honey and pollen.

Method of Application. The treatment is given in the evening, when the bees have ceased flying. Each hive should be prepared beforehand by adding below the crown board, an empty shallow super in which the smoke strip can be hung vertically from a small piece of wood or hardboard covering the feed-hole.

When all the bees have returned, close the hive entrance. Fix one end of a smoke strip to the underside of the piece of wood or hardboard with a drawing-pin and bend the strip so that it will hang vertically downwards. Light the other end of the strip, blow out the flame and place the wood or hardboard—with the glowing strip projecting into the space above the combs—over the feed-hole. If the crown board has two feed-holes, use the one in the centre and keep the other covered.

Leave the hive undisturbed for one hour; then open the entrance. The empty shallow super can be removed the next day.

If a box of super combs has recently been given to the colony, instead of adding an empty shallow box the super combs can be displaced temporarily to provide a gap below the feed-hole into which the strip can hang freely.

As an alternative to blocking the hive entrance and then releasing the bees an hour later, the entrance (or the narrow gap left when the usual type of cut-away entrance block is placed in position) may be closed by a strip of newspaper pinned across it. The bees will release themselves the next morning by biting their way through the newspaper.

One smoke strip at each application is adequate for colonies on a single brood chamber. Colonies on double brood chambers may be given two strips at each application.

EXAMINATION OF BEES FOR ACARINE

The nature of acarine, the ways in which it can be spread, and some methods of treatment, have already been described. The actual diagnosis can be made with certainty, however, only when the seat of infestation—the thoracic tracheae of the adult bee—by the mite *Acarapis woodi* Rennie has been exposed, following a dissection carried out with the aid of suitable instruments.

APPARATUS REQUIRED

A good dissecting lens, or a prismatic dissecting microscope giving an erect image;

A pair of finely pointed forceps;

A dissecting needle, preferably a double one in the form of a two-pronged fork, mounted in a handle;

An ordinary single dissecting needle, with handle;

A wide cork cut with a sloping surface and glued to a baseboard.

For comfort and ease of working a binocular dissecting microscope is undoubtedly better than a dissecting lens, but for most beekeepers the expense of such an instrument would not be justified, and satisfactory results can be obtained with a good aplanat giving a magnification of 8 or 10 diameters. A simple method of mounting a lens of this type is shown in Plate VII, with the forceps, needles and cork. The forceps, made preferably of fine sheet steel, should be a really good pair with a firm grip at the extreme tip of the blades. The double needle can be made by forcing the eyed ends of two No. 8 stainless sewing needles into a shaped handle of soft wood. The prongs of the fork should be parallel and $\frac{1}{16}$ in. apart. The cork should be about $2\frac{1}{2}$ in. diameter. The sloping surface can be cut with a hacksaw and smoothed with fine sandpaper.

Collecting Sample of Bees for Examination

For a periodic check on a colony, not less than 30 bees should be taken at random from the flight board, or from the feedhole in the crown board or quilt. They may be placed in match-boxes until the examination is made.

When disease is suspected, bees showing signs of sickness, collected from or near the flight board, form the best material for diagnostic purposes. Bees that have died recently are suitable for dissection but, generally speaking, those which have decomposed or dried up internally are useless.

Dissecting Bees

Any live bees in the sample should be killed, either by placing the match-box in an entomologist's killing bottle or by inserting into the match-box a piece of blotting paper soaked in chloroform. A bee is then taken out of the box, laid on its back with its head towards the right of the operator and gripped by the thorax with the forceps held in the right hand. The double needle, held in the left hand, is then thrust at an angle through the thorax, between the coxal joints of the second and third pairs of legs (Plate VII).

With the bee now firmly impaled on the needle, the forceps are released from the thorax and, while the needle is still held in the left hand, the point of one blade of the forceps is placed just behind the first pair of legs, and the point of the other blade is allowed to close on the head of the bee, midway between the antennae. The head and the first pair of legs are then detached cleanly from the bee, as one unit, by a sharp pull of the closed forceps towards the right and slightly upwards. Plate VII shows the parts to be removed at this stage and the plane of separation, as shown by the thick dotted line. The head and front pair of legs are discarded.

The remainder of the bee, still impaled on the needle, is now transferred to the sloping cork under the lens (Plate VIII). Looking through the lens, the focus is adjusted by moving the bee up or down the slope of the cork, until a clear view is obtained of the aperture in the thorax left by the removal of the head and first pair of legs (Plate VIII).

The chitinous 'collar' which surrounds the aperture (within the dotted line in Plate VIII) should now be removed to expose the thoracic tracheae on each side right back as far as the connection with the spiracle. The collar is gripped with the points of the forceps at X (Plate VIII), where it is very narrow, and peeled off with a rotary motion of the closed forceps in a clockwise direction. Some dexterity is needed to accomplish this, so that practice will be necessary before the collar can be taken off neatly with one sweep of the forceps. Very often a small triangular portion will be left at Y (Plate VIII), which should be removed separately.

The single mounted needle will be found useful for pushing aside tissues which may obscure a good view of the tracheae and for 'cleaning up' the dissection after the removal of the collar.

Fine adjustments of the focus can be made during the dissection by upward, downward, or twisting movements of the double needle holding the bee on the cork. The operation should be carried out in good light, preferably from a small spotlight or from a suitably screened reading lamp.

Diagnosis

The whole of that portion of the thoracic tracheal system which is liable to infestation by the mite now lies exposed (Plate VIII, b). The tracheae of a healthy bee have a uniformly smooth, creamy-white appearance (upper photograph, Plate IX). In contrast to this, a bee heavily infested with mites in both the left and right tracheal trunks is illustrated in the lower photograph, Plate IX. Note the blackening of the tracheae due to the activity of the mites within.

The difference between the two photographs of Plate IX is obvious, but intermediate stages between the two can also be recognized. Under a low-power lens the earliest stage which can readily be detected is marked by a bronzing of the tracheae along part of its length. This is followed by the appearance of black specks or streaks, which gradually enlarge and coalesce until finally the whole trachea may become uniformly black. The upper photograph of Plate X shows the thorax of a bee in which mites were present in the trachea on the right only, as indicated by the dark streaks and patches.

If a good binocular dissecting microscope, giving a magnification of 15 or 20 diameters, is used for the dissection, it is possible to detect the presence of individual mites before any discoloration of the trachea takes place. They appear as minute, oval, translucent bodies visible through the walls of the trachea. In order to see the mites clearly, however, it is necessary to transfer the trachea to a slide for examination under a compound microscope. Plate X (lower) is a photograph of a specimen mounted in a transparent medium showing mites in a heavily infested trachea.

When no signs of acarine are found in a sample consisting of sick or recently dead bees taken from or near the hive entrance, the cause of the trouble should be sought elsewhere; if a check on the stores in the hive shows that it was not due to starvation, another sample should be collected and sent away for examination (page 32). In a sample taken for examination as a routine precautionary measure, a lack of evidence of the presence of the mites does not necessarily mean that the colony is absolutely free from acarine, because in its earlier phases, when only a small number of bees in the colony will be infested, the chances of finding one or more infested bees in such a sample are correspondingly small. Periodic sampling for examination with each sample consisting of not less than 30 bees, will nevertheless enable the disease to be detected, and to be dealt with, before its effect on the colony become serious.

NOSEMA

The parasite *Nosema apis* Zander has long been known to occur in this country but it is only in relatively recent times that the attention of beekeepers has been drawn to its widespread incidence and to the serious effect it can sometimes have upon the productivity of their colonies. Much progress has now been made, however, towards a more complete understanding of the nature and spread of the infection and the new knowledge has enabled practical measures to be devised for the control of the parasite in the apiary.

THE NOSEMA PARASITE

The microscopic organism, *Nosema apis*, passes the active, reproductive phase of its life cycle within the digestive cells lining the mid-gut of the adult bee. Upon entering one of these cells the parasite grows and multiplies rapidly, utilizing the cell contents for its food supply, until reproduction ceases after a few days with the formation of a large number of spores. The host cell then ruptures, shedding the spores into the mid-gut, whence they pass down through the small intestine to the rectum. Here they accumulate in masses, to be voided at intervals in the excreta of the bee.

With the formation of spores the parasite enters into the passive or resting stage of its life cycle, from which it can emerge only if some of the spores from the excreta are picked up and swallowed by another bee. When this happens, the spores germinate on passing from the honey-stomach of the bee into the mid-gut. The parasites emerging from the spores then penetrate the lining of the mid-gut to start another phase of the intra-cellular growth and multiplication.

The length of time for which the spores can retain their ability to germinate depends upon the conditions to which they are exposed following their deposition with the excreta of infected bees. They remain viable for many months in dried spots of excreta on brood combs, for example, but lose their viability within a few days when suspended in water and exposed to direct sunlight. The spores are readily killed by heat and by suitable fumigants.

Worker bees infected with *Nosema apis* undergo a premature diminution in the size of the fat-body and the brood-food glands, and their average length of life is shorter than that of healthy bees. Where the infection is present, therefore, the brood-rearing and honey-producing capacity of a colony is reduced to an extent which will depend on the number of bees infected.

If the queen becomes infected her ovaries soon degenerate, and the result may be either a queenless colony or one in which the old queen is replaced by supersedure. Drones are not immune from infection by the parasite, but as they are normally present only during the active season when there is no transmission of the infection from bee to bee (see section on 'Spread' page 25), they are not an important factor in the course of the disease or in its effects upon the colony.

SYMPTOMS AND DIAGNOSIS

Bees infected with *Nosema apis* usually show no symptoms, or none that are specific for the infection. Inability to fly on leaving the hive, the dropping of excreta on the combs or on the alighting board, a heap of dead or moribund bees on the ground in front of the hive after a cleansing flight on a mild, sunny day in winter—these may be associated with nosema infection. But they may equally well be due to other abnormal conditions, such as starvation, the presence of fermenting stores in the combs, or a sudden drop in air temperature following a morning of warm, winter sunshine.

Usually when infection is present, no obvious signs of trouble are to be seen in or near the hive, but the colony may decline in strength owing to an abnormally high rate of loss of infected bees which die unnoticed away from the hive when foraging starts early in spring.

Only with the aid of the microscope can the infection be diagnosed with certainty. When infection is present the spores of *Nosema apis* (see Plate XI) can readily be detected in a watery suspension of the abdominal contents of the bees under examination, using a microscope giving a magnification of about 400 diameters. If, therefore, crawlers or unusual numbers of dead bees are seen in the apiary, or if a colony fails to build up properly in the spring, a sample of the bees should be sent for examination to the appropriate adult bee disease diagnosis centre (see Appendix page 32). Under the microscope the spores can also be detected in dry spots of excreta on soiled combs taken from infected colonies.

SPREAD

In order to understand how nosema infection is transmitted from bee to bee, it is best to consider an over-wintering colony with a few infected workers among the bees clustering on the combs.

Towards the end of the winter there is often some soiling of the combs with excreta dropped by both healthy and infected workers, though the colony as a whole may not exhibit any outward signs of a dysenteric condition. Infection spreads at once to any bees which lap up the newly-voided liquid excreta of infected bees and other bees will become infected later on, as they pick up the spores in the dried spots of excreta when cleaning the soiled combs during the spring expansion of the brood nest.

The level of infection within the colony rises quickly, and continues to rise for a time as more of the soiled comb area is cleaned to keep pace with the increasing amount of brood. The amount of brood will be less than it would be in a comparable healthy colony, however, owing to the depressing effect of the infection upon the efficiency of the workers as nurse bees and upon their length of life. The net result will be failure of severely infected colonies to develop normally in the spring. A colony in which a very high level of infection is reached during the winter may dwindle so rapidly in strength—owing to the premature death of the old, overwintered bees—that it is unable to survive.

Usually, however, colonies are not severely infected and they survive. The proportion of infected bees—after reaching its highest level in the late spring or early summer—quickly declines. The reason for this rapid decline is that, as soon as regular flights become possible in spring, the excreta are normally voided in the open air away from the hive. From this point onwards there is no further transmission of infection from the old, infected bees to the other occupants of an undisturbed hive, although young summer bees are inherently just as susceptible to infection as winter bees. The old bees die off and are replaced by the healthy bees emerging from the brood combs.

By the end of the season all or nearly all the old, infected bees will have gone. Nevertheless, enough spores remain on the combs from the previous winter to infect a few of the bees in the cluster which will form when winter sets in again. These infected bees will then form the starting point for a repetition of the whole cycle of events, leading to another peak of infection during the following spring.

Combs soiled by spore-containing excreta while the bees are confined to the hive during the winter months are, therefore, the natural carriers of nosema disease from year to year. The regular disappearance of the infection during the summer, when the bees are in frequent contact with their external environment, indicates that outside agencies—such as sources of drinking water, flowers or vegetation soiled by excreta—are not important factors in the spread of the disease. Honey is unlikely to be contaminated with the spores to any significant extent, since the deposition of excreta on the combs rarely occurs while the cells are being filled and sealed during the active season.

The spread of the disease from colony to colony—or apiary to apiary—can be brought about by the use of soiled brood combs, by the robbing of infected hives, and by the sale of nuclei, colonies, or queens and their attendant workers from an infected source.

Intensity of Infection

The intensity of the spring peak of infection varies from one year to another, often reaching a particularly high level when the preceding summer has been cool and wet, with a poor honey harvest.

Wet, cool conditions during the active season retard colony development; the combs are not fully occupied by brood or stores, and in consequence they are not cleaned by the bees as thoroughly as they are in a good season, when all the available cells are brought into use. The workers may live longer because of the smaller demands made upon them as nurses and foragers. The colony, therefore, reaches the end of the season with relatively 'dirtier' combs than usual, and possibly with some infected bees still present. The heavy autumn feeding which may be necessary will cause more bees to become infected while cleaning comb for the storage of the sugar syrup.

All these factors combine to provide the winter cluster with an abnormally large number of infected bees; the infection then reaches a correspondingly high peak of intensity during the following spring. Over a succession of bad summers the effect is likely to be cumulative.

Treatment

In considering control measures, other than those involving the use of drugs, the following features of the disease are of fundamental importance:

Infected bees cease to transmit infection to healthy bees during the flying season.

The natural carriers of the infection over the flying season are the combs on which bees clustered during the previous winter and early spring.

Any prolonged disturbance or excitement of colonies causing the deposition of excreta on the combs during the flying season will bring about an aggravation of infection at a time when it should be declining naturally as the old, infected bees die off.

Taking these features into account, an economic manipulative method of treatment becomes possible provided that the adult bees can be transferred on to *clean* combs early in the season without loss of brood or stores, and that the old, soiled combs can readily be made safe to use again. The following procedure satisfies both these conditions.

Manipulative Treatment

Transfer of Bees on to Full Set of Clean Combs. Early in the season, well before the main nectar flow, take a comb of brood with the queen on it, mark it, and place it in the centre of a clean brood box above the remaining combs, with a queen excluder between the two boxes. Fill up the top box with clean drawn combs that have been fumigated by the method to be described later. If clean drawn combs are not available, use new frames fitted with foundation. Feed with sugar syrup if there is any shortage of stores, or if frames of foundation are used instead of clean drawn comb.

About five days later, when the queen will be laying in one or more of the combs flanking the marked comb in the top box, transfer the marked comb back to the lower box, *leaving the queen up above the excluder*. When all the brood down below has emerged, removed the lower box, together with the floorboard and the queen excluder, and put the top box, with the queen and the new brood, down on a clean floorboard. During the three weeks while the brood is emerging below, the bees will transfer the stores from the lower box to the top box more readily if the floorboard entrance is closed, and a top entrance provided above the excluder.

If the old combs removed with the lower box are structurally sound, they should be fumigated as described later; otherwise they should be cut out of the frames and rendered down for wax, the temperature of molten beeswax being high enough to kill the nosema spores. The empty frames may be fumigated along with any sound combs.

The infection disappears as the old bees die off, and there will be no residual source of infection on the clean combs to start off another cycle of infection during the following winter.

Colonies should not be transported for long distances before about eight weeks have elapsed after their transfer to clean combs. Bees confined in hive or travelling box readily drop excreta when being moved from one site to another, and if this happens while any infected individuals are still present the combs will be recontaminated with nosema spores. The benefit of the transfer to clean combs will then be lost, and in addition the natural tendency of infection to decline will be halted, or even reversed. Similarly, care should be taken not to crush bees when replacing combs and hive parts after an examination of a treated colony.

Partial Replacement of Combs. The transfer of the bees on to a full set of clean combs is capable of eliminating the infection completely, by the removal of all the contamination which would have carried infection over the summer into the following winter. But the efficiency of the method may be upset by any subsequent, and perhaps unavoidable, movement or disturbance of the colony, as mentioned above.

An alternative procedure which involves less time and labour and reduces the amount of contamination is simply to remove as many as possible of the old combs on which the bees have wintered—leaving those which are occupied by brood—and to give in exchange clean, fumigated combs. A colony not covering a full set of combs when the old ones are removed is best contracted temporarily on to a smaller number which can be made up as the size of the colony increases.

Fumigation of Combs. Combs from infected colonies can be made safe to use again by exposing them to the fumes of 80 per cent *acetic acid.** The fumes kill the nosema spores but do not harm honey or pollen stored in the combs. The acid fumes have a corrosive effect upon metal objects, but the corrosion of nails, etc., in frames and hive parts is very superficial.

Where *metal-end spacers* are used, however, it is best to remove these from the frame lugs before fumigating the combs and to scald them in hot water containing washing soda or a household detergent compound.

Put the combs into deep or shallow boxes according to size, ten or eleven combs to a box with the usual spacing between them. Stack the boxes of combs on a floorboard *out of doors in a sheltered corner or in an open shed*, placing a wad of cotton wool or similar absorbent material—previously soaked in ¼ pint of the acid—in the bee space between each tier of combs, with another wad on top of the uppermost tier. (A 6-in. square of cheap grade cotton wool will readily take up the required amount of acid.)

Block the floorboard entrance, cover the top of the stack with a board or hive roof and seal any obvious gaps in the stack through which the fumes might escape; then leave the stack undisturbed for one week. The fumes act on the spores more quickly in warm weather, but a week is long enough even under the coldest conditions likely to be encountered in this country.

The acid should be handled with care to avoid splashing the liquid on the face or into the eyes. Splashes on the hands should be washed off without delay.

After fumigation the combs should be aired for about 48 hours before they are used again. Take the board or hive roof off the top, remove all the wads and unblock the floorboard entrance. If any of the combs contain honey they may be aired under cover as a precaution against robbing; alternatively, the stack of boxes may be protected top and bottom by bee-proof screens.

Fumigation with acetic acid will also control any wax-moth infestation which may be present in the combs. The eggs and adults of both the greater and the lesser wax-moth are killed within seven days by the acid fumes, also all their larval stages—except for the last larval stage of the greater wax-moth. This, too, can be killed if paradichlorbenzene (PDB) is used in conjunction with the acetic acid. A handful of PDB crystals should be

*One part by volume of water to four parts of *glacial* acetic acid (obtainable from a chemist). Larger quantities of a suitable low-grade 80 per cent acetic acid are available from trade suppliers of industrial chemicals.

placed on a piece of paper laid across the top bars of the uppermost tier of frames in the stack of boxes prepared for fumigation. The combs should be aired after fumigation until the smell of PDB is no longer perceptible.

Drug Treatment

The infection of the mid-gut can be arrested, though not always entirely cured, if the infected bees have continuous access to a supply of sugar syrup containing the antibiotic drug *fumagillin*. But the infection cannot be completely eliminated from the colony by this treatment alone; the drug has no effect upon the spore stage of the nosema parasite and, when all the medicated syrup has been consumed, enough spores usually remain on the combs to renew the infection later.

Fumagillin is not recommended for use simultaneously with the manipulative method of control described above, which takes advantage of the natural decline of infection as the old infected bees die off during the active season. The best time to use fumagillin is in the autumn, either as a preliminary to a transfer to clean combs in the spring, or as an additional precautionary measure at the end of the season following the transfer. The drug can readily be incorporated in the syrup used for feeding in the usual way.

Feeding with syrup containing fumagillin very early in the season, before the bees are flying regularly, may help to check the spread of the disease within the colony at this time of year. But any permanent benefit from it will be lost unless it is followed by the manipulative method of control, and the removal of the old combs for fumigation before they are used again.

A commercial preparation containing fumagillin is now available. Full instructions for use are supplied with it by the makers.

AMOEBA

'Amoeba' is the term commonly used to denote the invasion of the excretory organs (Malpighian tubules) of the bee by a microscopic organism known as *Malpighamoeba mellificae* Prell. The parasites develop and multiply at the expense of the excretory cells lining the tubules and then form spherical cysts which pass down through the small intestine to the rectum. In the rectum they accumulate and are voided from it at intervals in the excreta of the bee. Cysts which are picked up and swallowed by another bee with its food complete their life cycle by entering into another phase of active growth and multiplication in the tubules of their new host. The time taken from the initial infection of the bee to the formation of the cysts is from 24 to 28 days.

Diagnosis of infection depends upon the discovery of the cysts under the microscope, either in samples of bees or in dried spots of excreta on combs fouled by infected bees. There are no specific symptoms and infected bees usually appear normal.

No form of drug treatment is known which will effect a cure of infected bees, but the infection can be eliminated from a colony by the manipulation method already described for a nosema infection (page 27). Similarly, combs taken from infected colonies can be made safe to use again by the acetic acid fumigation method which is effective against nosema spores.

BEE PARALYSIS

The chronic form of 'paralysis' among bees is that which is known to be caused by a virus. It is characterized by the presence of bees unable to fly, often having trembling or shaky movements of the legs and wings, and sometimes with a hairless and polished or greasy appearance of the thorax and abdomen.

Bees having these symptoms are to be seen at or near the hive entrance, where up to many hundred specimens may be collected daily from a colony that is badly affected. The other bees of the colony use their mandibles to push or nibble at the paralysed bees and this behaviour on their part is probably the cause of the hairless, shiny appearance of many of the sick bees.

Many, probably most, colonies have some infected bees but few colonies are very susceptible to the virus, and where susceptibility does occur it seems to be limited to the progeny of particular queens. There is, therefore, virtually no risk of the active disease being spread to other colonies by the transfer of bees, brood or equipment from a diseased colony. Re-queening of the affected colony is advised as the most likely means of overcoming the trouble. No method of treating an infection already present in a colony is known, but when a new queen has been accepted the signs of the disease should disappear as the progeny of the old queen die off.

Other forms of 'paralysis', transient rather than chronic in their effect on a colony, can be caused by natural poisons present in a variety of sources of nectar, honeydew and pollen. Among the factors known or suspected to be the cause of such sickness are the collection of pollen from some species of buttercup; the collection of nectar from some species of rhododendron and two species of lime (*Tilia petiolaris* and *T. orbicularis*); and of honeydew from trees, including limes and oaks. These forms of paralysis disappear after a short time as the bees turn their attention to other sources of forage and are less damaging than the chronic virus paralysis, which can cause losses of bees over a long period and lead to the severe weakening or even the complete loss of the colony.

DYSENTERY

The term 'dysentery', as used by beekeepers, refers not to any infectious disease but to the involuntary discharge of excreta by the adult bees within and near the hive in the late winter or early spring, so that the combs and hive entrance become visibly soiled with faeces. Normally bees discharge their faeces in flight away from the hive and do not soil their combs or the exterior parts of the hive.

A dysenteric condition may arise simply from an unusually prolonged period of confinement to the hive during a severe winter giving no opportunities for cleansing flights. More usually, however, it is caused by the consumption of unsuitable food during the winter—acid-inverted invert sugar, for example, or honey which is too watery. The condition may prove lethal to a colony when most of its members are affected simultaneously; the watery faecal matter is consumed by the bees and conditions within the hive rapidly become worse; the bees soon die of suffocation from being smeared with the excrement and from the unusual turmoil that arises from this crisis in the colony's behaviour.

Though dysentery is not caused by an infectious disease, any infection with nosema or amoeba disease will clearly be aggravated by the onset of a dysenteric condition, since the contaminated faeces are consumed by the bees, which are thus directly exposed to further infection.

Exceptionally severe and prolonged winters apart, trouble from dysentery should not normally be experienced if colonies are well provisioned with suitable stores in the autumn. Honey which crystallizes coarsely in the combs is unsuitable, since the bees are unable to utilise the coarse crystals, and the surrounding liquid, now contains all the water in the original honey; this excess water is a direct cause of dysentery, apart from making the honey liable to ferment. Strong sugar syrup, made from refined white sugar and fed before the end of September, produces stores which are much less likely to crystallize than some types of honey and which are therefore very suitable as winter food for the bees.

Appendix

SUBMISSION OF SAMPLES OF BROOD OR ADULT BEES FOR DISEASE DIAGNOSIS

Brood Combs

Sample brood combs for laboratory examination from beekeepers in England should be sent to the Beekeeping Adviser, Agricultural Development and Advisory Service, Rothamsted Lodge, Hatching Green, Harpenden, Hertfordshire; beekeepers in Wales should address samples to the Beekeeping Adviser, Agricultural Development and Advisory Service, Trawscoed, Aberystwyth, Cardiganshire.

Take one of the combs showing the signs of suspected disease, preferably one containing brood in all stages but without a lot of honey, and carefully shake the adhering bees back into the hive. Remove metal end spacers and break off the lugs of the frame. Then wrap the comb in several thicknesses of newspaper and pack it securely in a strong cardboard box, placing the covering note outside the inner newspaper wrappings and not in direct contact with the comb.

Always send a complete comb if possible, not individual larvae removed from their cells. If a piece of comb from a box hive or skep is sent as a sample, do not use a tin box or a polythene bag as a container; the 'sweating' of the sample in an enclosed space with no ventilation favours the rapid decomposition of the brood, particularly in hot weather, and the original condition of the brood may be quite unrecognizable by the time the package is opened in the laboratory.

Sample combs taken by Appointed Officers under the Foul Brood Disease of Bees Order are packed in the special wrappings and boxes with which the officers are provided for sending combs through the post.

Adult Bees

A service for the diagnosis of disease in adult bees is now provided in a number of counties where County Beekeeping Instructors have been trained and equipped for this work. Lists giving the addresses to which bees may be sent for local examination are published from time to time in the beekeeping journals. If no local service is available samples should be sent to the Agricultural Development and Advisory Service at one of the addresses given in the preceding section on sending samples of brood.

Each sample should consist of not less than 30 bees, all taken from the same hive, placed in a match-box or similar container (not in a tin box or a polythene bag) clearly labelled with the name and address of the sender and the number or other identification mark of the hive concerned. Relevant information about the sample should be given in a covering note enclosed with the package.

Old bees are best for laboratory examination. Very young bees from the face of a comb of brood should not be taken. One of the quickest and simplest methods of securing a sample is to scoop the bees into an inverted match-box tray, keeping them trapped under the tray whilst the cover is slid on. This method can be used on any flat surface covered with bees, such as the underside of a crown board, along the top bars of the frames or the face of a comb. A useful item of equipment is a small square of glass under which the match-box containing a sample can be opened for inspection, without releasing the bees, to estimate how many have been captured.

Losses of Adult Bees not due to infectious disease

When a sample of bees is received for examination the bees are first dissected for acarine and then tested for nosema and amoeba. If no evidence of any of these is found under the microscope, and if neither the covering letter nor the appearance of the bees suggests a form of 'paralysis', the cause of the trouble in the bees comprising the sample must be something other than disease infection. The following notes indicate some of the more common causes of losses of adult bees where no evidence of disease can be found.

Starvation. If the whole colony is found to be dead, with many of the bees occupying empty cells, each lying head first and with only the end of the abdomen showing, the trouble is almost certainly due to starvation. Sometimes a small cluster of dead bees is found on combs containing, or adjacent to, sealed cells full of stores. The probably explanation here is that the colony died out because it was too small to maintain a degree of warmth sufficient to enable the cluster to move within reach of the food still present in the combs.

A condition of near-starvation, during a period of cold or wet weather early in the season before the main nectar-flow, is often characterized by the appearance of dead drones and worker pupae on or near the alighting board of the hive.

Chilling. A sudden fall in air temperature following a warm, sunny midday in early spring may result in the formation of small clusters of apparently lifeless bees on the ground near the entrance, or underneath the floorboard. This may be due to the bees having become chilled before they were able to enter the hive on their return from a flight in search of forage or water.

Robbing. Casualties to bees defending their hives from robbing in the autumn, either by other bees or sometimes by wasps, may be quite heavy. If a weak colony is robbed out completely the empty combs will have small particles of wax sticking to them where the cappings have been torn away by the robbers. Fragments of cappings will be found scattered on the floorboard.

Spray poisoning. If there is a sudden and serious reduction in the number of flying bees in an apiary, perhaps accompanied by crawlers or clusters of dead or dying bees near the hives, the reason may be that the foragers have been poisoned while working on a crop which has been sprayed with insecticide or weed-killer. Local enquiries should be made at once to discover whether the start of the trouble coincided with the spraying or dusting of orchard or field crops in the locality.

If a sample of the dead bees is sent for examination it should consist of *at least* 200 *bees* (one match-box of the usual size will hold about 50 dead bees). It should be accompanied by information about the extent of the damage (for example, how many colonies in the apiary have been affected, whether similar trouble has been experienced by neighbouring beekeepers), about spraying or dusting operations known to have been carried out at or about the time when the trouble was first noticed, including the type of crop and, in particular, the type of spray or dust used on the crop.

Printed in Scotland by Her Majesty's Stationery Office at HMSO Press, Edinburgh
Dd 696889 K10 8/80 (17493)